CLIMATE CHANGE SCIENCE PROGRAM and
SUBCOMMITTEE ON GLOBAL CHANGE RESEARCH

William J. Brennan
Department of Commerce
National Oceanic and Atmospheric Administration
Director, Climate Change Science Program; and
Chair, Subcommittee on Global Change Research

Jack Kaye, Vice Chair
National Aeronautics and Space
Administration

Thomas Armstrong
U.S. Geological Survey

Allen Dearry
Department of Health and Human Services

Mary Glackin
Department of Commerce

Patricia Gruber
Department of Defense

William Hohenstein
Department of Agriculture

Timothy Killeen
National Science Foundation

Linda Lawson
Department of Transportation

Patrick Neale
Smithsonian Institution

Anna Palmisano
Department of Energy

Jacqueline Schafer
U.S. Agency for International Development

Joel Scheraga
Environmental Protection Agency

Harlan Watson
Department of State

EXECUTIVE OFFICE AND OTHER LIAISONS

George Banks
Council on Environmental Quality

Howard Frumkin
Centers for Disease Control and Prevention

Katharine Gebbie
National Institute of Standards and Technology

Stuart Levenbach
Office of Management and Budget

Robert Marlay
Department of Energy
Climate Change Technology Program

Margaret R. McCalla
Office of the Federal Coordinator for Meteorology

Daniel Walker
Office of Science and Technology Policy

OUR CHANGING PLANET

The U.S. Climate Change Science Program
for Fiscal Year 2009

A Report by
the Climate Change Science Program and
the Subcommittee on Global Change Research

A Supplement to the President's Budget for Fiscal Year 2009

July 2008

Members of Congress:

We are pleased to transmit a copy of *Our Changing Planet: The U.S. Climate Change Science Program for Fiscal Year 2009*. The report describes the activities and plans of the Climate Change Science Program (CCSP), which incorporates the U.S. Global Change Research Program established under the Global Change Research Act of 1990, and the Climate Change Research Initiative that was established by the President in 2001. CCSP coordinates and integrates scientific research on climate and global change supported by 13 participating departments and agencies of the U.S. Government.

This Fiscal Year 2009 edition of *Our Changing Planet* highlights recent advances and progress supported by CCSP-participating agencies in each of the program's research and observational elements, as called for in the *Strategic Plan for the U.S. Climate Change Science Program* released in July 2003, and later modified in the 2008 CCSP *Revised Research Plan*.

The document describes a wide range of activities including examples of CCSP's contribution to the Fourth Assessment Report of the Intergovernmental Panel on Climate Change as well as significant progress in understanding Earth system components of the global climate system, how these components interact, and the processes and forces bringing about changes to the Earth system. It provides details on progress towards understanding the ongoing and projected effects of climate change on nature and society, such as the relationship between climate change and shifts in storm tracks and how this may affect water availability in the southwestern United States. The document also describes how observational and predictive capabilities are being improved and used to create tools to support decisionmaking at local, regional, and national scales to cope with environmental variability and change. The report also describes the program's 21 scientific synthesis and assessment products and its recently completed *Scientific Assessment of the Effects of Global Change on the United States*. To date, eight of the synthesis and assessment products have been completed and the remaining products will be completed in the coming months. These products are being widely disseminated and briefed to stakeholders, including Congress. They are also providing important input to CCSP's ongoing strategic planning.

CCSP is committed to its mission to facilitate the creation and application of knowledge of the Earth's global environment through research, observations, decision support, and communication. We thank the CCSP-participating agencies for their close cooperation, and we look forward to working with Congress in the continued development of this important program.

Carlos M. Gutierrez

Secretary of Commerce

Chair, Committee on Climate Change Science and Technology Integration

Samuel W. Bodman

Secretary of Energy

Vice Chair, Committee on Climate Change Science and Technology Integration

John H. Marburger III

Director, Office of Science and Technology Policy

Executive Director, Committee on Climate Change Science and Technology Integration

TABLE OF CONTENTS

THE U.S. CLIMATE CHANGE SCIENCE PROGRAM FOR FY 2009

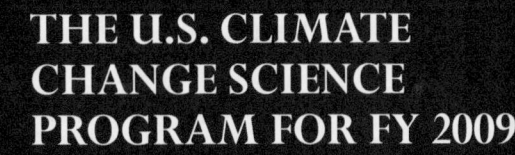

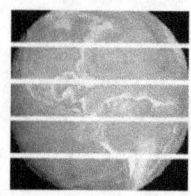

THE U.S. CLIMATE CHANGE SCIENCE PROGRAM FOR FY 2009

Climate plays an important role in shaping the environment, natural resources, infrastructure, economy, and other aspects of life in all countries of the world. Therefore, variations and changes in climate can have substantial environmental and socioeconomic implications. The Climate Change Science Program (CCSP) was established in 2002 to empower the Nation and the global community with the science-based knowledge to manage risks and opportunities of change in the climate and related environmental systems. CCSP incorporates and integrates the U.S. Global Change Research Program (USGCRP) with the Administration's U.S. Climate Change Research Initiative (CCRI). The USGCRP was mandated by Congress in the Global Change Research Act of 1990 (P.L. 101-606, 104 Stat. 3096-3104) to improve understanding of uncertainties in climate science, expand global observing systems, develop science-based resources to support policymaking and resource management, and communicate findings broadly among scientific and stakeholder communities.

Climate research conducted over the past several years indicates that most of the global warming experienced in the past few decades is very likely due to the observed increase in greenhouse gas concentrations from human activities. Research also indicates that the human influence on the climate system is expected to increase.[1] It is therefore essential for society to be equipped with the best possible knowledge of climate variability and change so that it may exercise responsible stewardship for the environment, lessen the potential for negative climate impacts, and take advantage of positive opportunities where they exist. The importance of these issues and the unique role that science can play in informing society's responses give rise to CCSP's guiding vision.

CCSP GUIDING VISION

A nation and the global community empowered with the science-based knowledge to manage the risks and opportunities of change in the climate and related environmental systems.

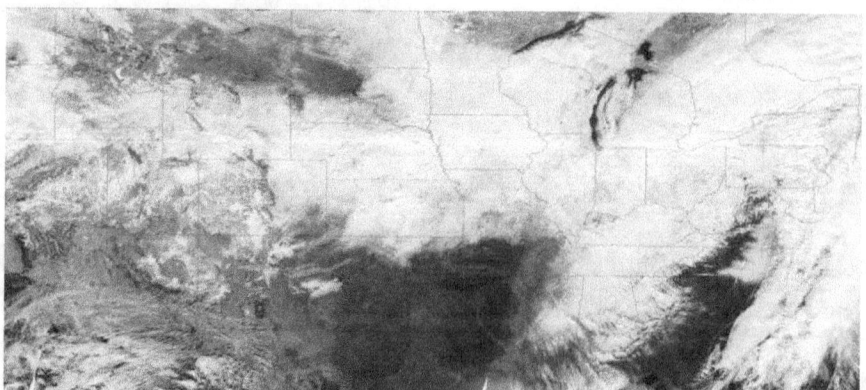

CCSP carries out its mission through four core approaches: scientific research, observations, decision support, and communication. These approaches build upon scientific advances of the last few decades and are deepening understanding of the interplay of natural and human-caused forces, their implications, and response options. CCSP is developing information to facilitate comparative analysis of different approaches for adapting to and mitigating climate change. CCSP also promotes capacity development among scientists and information users—both in the developed and the developing world—to address the interactions between climate change, society, and the environment.

> ### CCSP MISSION
>
> **Facilitate the creation and application of knowledge of the Earth's global environment through research, observations, decision support, and communication.**

INTEGRATING CLIMATE AND GLOBAL CHANGE RESEARCH

Thirteen departments and agencies of the U.S. Government participate in CCSP, including:
- Department of Agriculture (USDA)
- Department of Commerce (DOC)
  - National Oceanic and Atmospheric Administration (NOAA)
  - National Institute of Standards and Technology (NIST)
- Department of Defense (DOD)
- Department of Energy (DOE)
- Department of Health and Human Services (HHS)
- Department of the Interior/U.S. Geological Survey (DOI/USGS)
- Department of State (DOS)
- Department of Transportation (DOT)
- Agency for International Development (USAID)
- Environmental Protection Agency (EPA)
- National Aeronautics and Space Administration (NASA)
- National Science Foundation (NSF)
- Smithsonian Institution (SI).

In addition, the Executive Office of the President and other related programs have designated liaisons who participate on the CCSP Interagency Committee, including:
- Office of Science and Technology Policy (OSTP)
- Council on Environmental Quality (CEQ)
- Office of Management and Budget (OMB)

- Climate Change Technology Program (CCTP)
- Office of the Federal Coordinator for Meteorology (OFCM).

Appendix A, "The Climate Change Science Program Participating Agencies," contains information about the specific missions and roles of each agency participating in CCSP. Appendix B, "Climate Change Science Program FY 2008 Budget Tables," contains budgetary analyses of the program grouped by agency as well as a program-wide interagency cross-cut grouped by the strategic goals and research elements of CCSP as described in the *Strategic Plan for the U.S. Climate Change Research Program* published in July 2003.

CCSP is responsible for coordinating and integrating scientific research on global environmental variability and change sponsored by these agencies to take advantage of their unique approaches and missions, and to encourage research that leads to expanded and new results. Thus, the program helps to catalyze research that goes beyond individual agency missions to address overarching national objectives and to achieve results that no single agency, or small group of agencies, could attain. A significant challenge that arises from working across many agencies is integrating climate and global change research to develop a comprehensive view of climate change and its potential significance.

CCSP relies not only on the agency programs stated in its budget cross-cut, but also on agency activities that are not formally included in the CCSP budget. Examples of these directly related activities are surface hydrologic and satellite land-cover observations from USGS; and future satellite measurement programs including the tri-agency (NOAA, DOD, NASA) National Polar-Orbiting Operational Environmental Satellite System (NPOESS) and the planned implementation of a Landsat Data Continuity Mission (LDCM).[a] Without input from activities such as these, CCSP would be unable to fulfill its mission.

CCSP is closely allied with other major interagency programs that observe and study particular aspects of the Earth system and related societal dimensions. Foremost among these is the CCTP, which develops and studies technological options for responding to climate change. CCSP is also closely linked to ongoing Federal ocean science and technology strategic planning under the auspices of the Joint Subcommittee on Ocean Science and Technology, which recently released a set of integrating decadal-scale national research priorities in key areas of interaction between society and the ocean.[2] A key observational linkage is with the U.S. Integrated Earth Observation

---

[a] As a result of the recent review and reformulation of its CCSP contributions, NASA considers 33% of its LDCM budget and 100% of its NPOESS Preparatory Project budget to contribute to CCSP. NASA budget figures provided in this report reflect the inclusion of this funding.

DEFINITION OF KEY TERMS

Adaptation
Adjustment in natural or human systems to a new or changing environment that exploits beneficial opportunities or moderates negative effects.

Climate
The statistical description of the mean and variability of relevant measures of the atmosphere-ocean system over periods of time ranging from weeks to thousands or millions of years.

Climate Change
A statistically significant variation in either the mean state of the climate or in its variability, persisting for an extended period (typically decades or longer). Climate change may be due to natural internal processes or to external forcing, including changes in solar radiation and volcanic eruptions, or persistent human-induced changes in atmospheric composition or in land use.

Climate Feedback
An interaction among processes in the climate system in which a change in one process triggers a secondary process that influences the first one. A positive feedback intensifies the change in the original process, and a negative feedback reduces it.

Climate Forcing
A process that directly changes the average energy balance of the Earth-atmosphere system by affecting the balance between incoming solar radiation and outgoing or "back" radiation. A positive forcing tends to warm the surface of the Earth and a negative forcing tends to cool the surface.

Climate System
The highly complex system consisting of five major components: the atmosphere, the hydrosphere, the cryosphere, the land surface, the biosphere, and the interactions between them. The climate system evolves in time under the influence of its own internal dynamics and because of external forcings such as volcanic eruptions, solar variations, and human-induced forcings such as the changing composition of the atmosphere and land-use change.

Climate Variability
Variations in the mean state and other statistics of climatic features on temporal and spatial scales beyond those of individual weather events. These often are due to internal processes within the climate system. Examples of cyclical forms of climate variability include the El Niño Southern Oscillation, the North Atlantic Oscillation, and the Pacific Decadal Oscillation.

Decision-Support Resources
The set of observations, analyses, interdisciplinary research products, communication mechanisms, and operational services that provide timely and useful information to address questions confronting policymakers, resource managers, and other users.

Global Change
Changes in the global environment (including alterations in climate, land productivity, oceans or other water resources, atmospheric chemistry, and ecological systems) that may alter the capacity of the Earth to sustain life (from the Global Change Research Act of 1990, PL 101-606).

Mitigation
An intervention to reduce the human-induced factors that contribute to climate change. This could include approaches devised to reduce emissions of greenhouse gases to the atmosphere; to enhance their removal from the atmosphere through storage in geological formations, soils, biomass, or the ocean; or to alter incoming solar radiation through several "geo-engineering" options.

Observations
Standardized measurements (either continuing or episodic) of variables in climate and related systems.

Prediction
A probabilistic description or forecast of a future climate outcome based on observations of past and current oceanic and atmospheric conditions and quantitative models of climate processes (e.g., a prediction of an El Niño event).

Projection
A description of the response of the climate system to an assumed level of future radiative forcing. Changes in radiative forcing may be due to either natural sources (e.g., volcanic emissions) or human-induced factors (e.g., emissions of greenhouse gases and aerosols, or changes in land use and land cover). Climate "projections" are distinguished from climate "predictions" in order to emphasize that climate projections depend on scenarios of future socioeconomic, technological, and policy developments that may or may not be realized.

Weather
The specific condition of the atmosphere at a particular place and time, measured in terms of variables such as wind, temperature, humidity, atmospheric pressure, cloudiness, and precipitation.

System, which is part of the international Global Earth Observation System of Systems (GEOSS). Connections to programs such as these allow CCSP and its partners to leverage their resources to derive mutual benefits from advances in any one program.

In May 2008, CCSP released a *Revised Research Plan for the U.S. Climate Change Science Program*. This updated plan is required in order to reflect both scientific advances since the publication of CCSP's 2003 Strategic Plan, and the evolving needs of society. It provides direction for addressing remaining uncertainties in climate science, including impacts at regional scales and adaptation options. The Revised Research Plan also emphasizes the need for strengthened communication of scientific studies to decisionmakers across the United States. The updated plan focuses on near-term (1-3 year) planning needs, and specifically addresses research plans for the period 2008-2010. The release of the plan is part of ongoing extensive strategic planning in the Climate Change Science Program that began in 2007. The Revised Research Plan addresses the research plan requirements in Sec. 104(a) of the Global Change Research Act of 1990.

Program Management

CCSP's coordination of scientific research is accomplished through the research elements described in the following section. The management approach as described in the CCSP Strategic Plan integrates the planning and implementation of individual climate and global change research programs of the participating Federal agencies and departments to reduce overlaps, identify and fill programmatic gaps, and synthesize products and deliverables generated under the auspices of CCSP. Five mechanisms are used to achieve this management approach:

- *Executive Direction* – The Interagency Working Group on Climate Change Science and Technology and the CCSP Interagency Committee are responsible for overall priority setting, program direction, management review, and accountability to deliver program goals.
- *Agency Implementation* – CCSP-participating departments and agencies are responsible for conducting research, developing modeling tools, developing and operating observing systems, and producing CCSP-required products, often in collaboration with interagency working groups.
- *Interagency Planning and Implementation* – Several interagency working groups, including one for each CCSP research element, are responsible for coordinating planning and implementation to align agency programs with CCSP priorities.
- *External Guidance and Interaction* – External advisory groups and organizations, including the National Academies, provide external guidance, oversight, and interactions to ensure scientific excellence, credibility, and utility.

- *Program Support* — The CCSP Office provides staffing and day-to-day coordination of CCSP-wide program integration, strategic planning, product development, and communications.

Coordinating Research Elements

Efforts to foster integration occur on many levels. One is improving coordination of scientific research and the flow of information through interdisciplinary and interagency working groups focused on each of seven main research elements of the program plus a number of cross-cutting activities or themes. CCSP's research elements include atmospheric composition, climate variability and change, the global water cycle, land-use and land-cover change, the global carbon cycle, ecosystems, and human contributions and responses to environmental change. Chapters 3 to 15 of the CCSP Strategic Plan contain more detailed discussions of the discipline-specific research elements, as well as elements that cut across all areas of the program. A brief summary of each of these research and cross-cutting elements is provided herein, as well as a few highlights of planned activities.

Integrating research and observational approaches across disciplinary boundaries is essential for understanding how the Earth system functions and how it will change in response to future forcing. This is due to the interconnectedness among components of the Earth system, which often relate to each other through feedback loops. Interdisciplinary interactions in CCSP are scaled to the nature of the problem. In some cases, the necessary science may be conducted within a small set of disciplines, such as those required to improve understanding of soil biogeochemical processes. In other

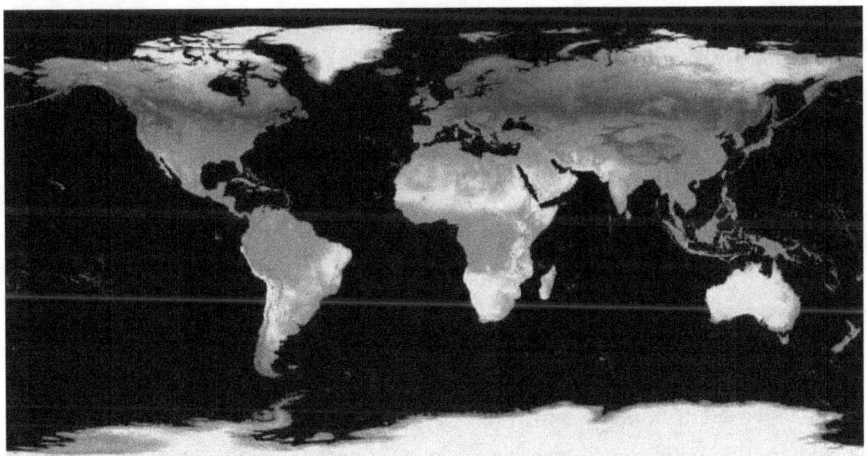

cases, highly interdisciplinary and multi-scale approaches are required, such as in the case of making projections about the future state of the Earth system and analyzing their implications. In this case, expertise ranging from the social sciences to atmospheric dynamics and chemistry to oceanography to the biological sciences is required. Examples of interdisciplinary research are the coordinated planning and operation of two intensive field experiments in the Arctic region and in the tropics. The first concerned the climate impacts of aerosols and clouds in the Arctic as part of the International Polar Year (see <www.polarcat.no>). The second focused on expanding the scientific understanding of climate-cloud-chemistry interactions in the highly active and climate-relevant region of the tropical atmosphere (see <www.espo.nasa.gov/tc4>). These campaigns, involving several different agencies, were designed to address interdisciplinary science questions spanning three CCSP research elements—global water cycle, atmospheric composition, and global carbon cycle.

Interdisciplinary research is only one aspect of the integration facilitated by CCSP. Integration in CCSP also refers to the steps being taken to create more seamless approaches between the theory, modeling, observations, and applications that are required to address the multiple scientific challenges being confronted by CCSP. Finally, integration in CCSP also refers to the enhancement of cooperation across agencies toward meeting the objectives articulated in the CCSP Strategic Plan.

Integrated Program Analysis

In a highly distributed program such as CCSP, it is often a challenge to develop and maintain a cohesive perspective, ensuring that key components or interactions of the integrated Earth system are not overlooked. To help address this challenge, the program has often sought guidance from the National Academies. CCSP is funding a National Research Council (NRC) committee to provide high-level, independent, integrated advice on the strategy and evolution of the program. Specific topics follow:
- The first committee report included findings and recommendations on the process for evaluating progress toward the five CCSP goals and a preliminary assessment of progress to date.
- The second report will identify priorities to guide the future evolution of the program in the context of established scientific and societal objectives.

At the request of CCSP, the NRC recently produced a report on global change assessments that is briefly described in a later section.

CCSP will continue to rely on other mechanisms for scientific guidance and advice, including other NRC committees that focus on particular components of the climate system (e.g., the Climate Research Committee and the Committee on the Human Dimensions of Global Change). CCSP will continue to rely on scientific advisory groups that support individual agencies, scientific steering groups organized to coordinate different CCSP research elements, and open dialog with the domestic and international scientific and user communities interested in global change issues.

CCSP GOALS AND ANALYSIS OF PROGRESS TOWARD THESE GOALS

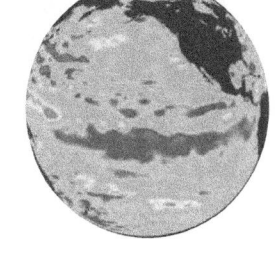

The Climate Change Science Program focuses on five goals that address the full range of global change research, observations, and decision support. These goals address understanding the components of the Earth's varying environmental system, with a particular focus on climate; understanding how these components interact to determine present conditions; understanding what drives these components; understanding the history of global change and projecting future change; and understanding how knowledge about global environmental variability and change can be applied to present-day and future decisionmaking.

This section provides an overview of the progress made toward achieving these goals in the 12 to 18 months prior to the preparation of the FY 2009 edition of *Our Changing Planet*. Because of the breadth of climate research funded by the U.S. Government, this overview only provides a general summary of some of the many climate change research activities covered under the CCSP umbrella.

In the past decade, the primary focus of U.S. climate research has been on CCSP Goals 1, 2, and 3—emphasizing understanding the global climate system through observations, identifying the various components of the global climate system, understanding how

---

CCSP GOALS

Goal 1: Improve knowledge of the Earth's past and present climate and environment, including its natural variability, and improve understanding of the causes of observed variability and change.

Goal 2: Improve quantification of the forces bringing about changes in the Earth's climate and related systems.

Goal 3: Reduce uncertainty in projections of how the Earth's climate and related systems may change in the future.

Goal 4: Understand the sensitivity and adaptability of different natural and managed ecosystems and human systems to climate and related global changes.

Goal 5: Explore the uses and identify the limits of evolving knowledge to manage risks and opportunities related to climate variability and change.

the various components interact to drive the climate system, and working toward developing predictive tools that identify near- and long-term climate variability. As progress towards a better understanding of the global climate system continues to be made, Goals 4 and 5 are gaining additional attention.

The following are recent examples of progress contributing to each of these five goals, resulting from coordinated research activities in many disciplines conducted by or supported across the CCSP-participating agencies. In all of these areas, CCSP has functioned to facilitate interagency cooperation and coordination within the U.S. Government.

The FY 2009 edition of *Our Changing Planet* summarizes a massive array of scientific evidence that human activities are responsible for recent global warming. Recent studies have (1) documented and corrected errors in several previous studies that cast doubt on global warming, (2) reinforced the conclusion that solar forcing is not responsible for global warming for the past few decades, and (3) further documented ongoing global warming via observations of many different facets of the global climate system. Taken together, these recent reports provide unequivocal evidence that the planet is warming, and further research provides coherent evidence that it is very likely caused by human activities. These conclusions by the Intergovernmental Panel on Climate Change (IPCC), made possible in part through CCSP research (see box on pages 11 and 12), emphasize CCSP's urgent mission of interagency cooperation and coordination. In order to continue to improve understanding of ongoing and future changes, in particular to provide information to support decisionmaking, the need for sustained satellite and *in situ* climate observations is underscored. These observational activities must be coupled with robust modeling and analysis, as well as research on impacts, adaptation, and vulnerability to global environmental variability and change.

CCSP's series of synthesis and assessment products, briefly described in this chapter, summarize the current state of interagency global change research within a multi-agency context. These products span all five CCSP goals. With their in-depth focus on North America, they provide an important complement to the IPCC reports.

In May 2008, CCSP released a report entitled *Scientific Assessment of the Effects of Global Change on the United States*, which draws from the IPCC reports and several of the synthesis and assessment products. The scientific assessment integrates, evaluates, and interprets the findings of CCSP and other scientific investigations of global change. It analyzes the trends and effects of global environmental changes, with a particular focus on climate change impacts on the United States. Together with CCSP's synthesis and assessment products and other major assessments, the report is an important tool for

EXAMPLES OF CCSP CONTRIBUTIONS
TO THE IPCC FOURTH ASSESSMENT REPORT

CCSP scientists in cooperation with other international scientists contributed significantly to the Fourth Assessment Report of the IPCC, published in 2007. One of the highlights of this cooperation was the development of a large set of coupled atmosphere-ocean global climate model experiments for 20th and 21st century climate. This effort, as well as the subsequent analysis phase, was organized through CCSP-sponsored cooperation with the World Climate Research Programme (WCRP) and its Climate Variability and Predictability (CLIVAR) project. The resulting data set is the largest and most comprehensive international global coupled climate model experiment and multi-model analysis effort ever attempted, and is openly available for analysis and academic applications. Over 200 journal articles, based in part on the data set, have been published or submitted for publication to date, and many more are being prepared. The ready access to the multi-model data set opens up these types of model analyses to researchers, including students and developing country scientists, who previously could not obtain state-of-the-art climate model output, and thus represents a new era in climate change research. As a direct consequence, these ongoing studies are increasing the body of knowledge regarding understanding of how the climate system currently works, and how it may change in the future.

With wide public acceptance of the IPCC Fourth Assessment Report findings, the climate science community is moving beyond simply running the basic scenarios to prove that climate change is occurring into new areas of focused climate change decision support. This will involve putting climate change in the context of the Earth's evolving environmental systems to assess the adaptation and mitigation implications of local, national, and international policies on emissions, energy use, and resource management. Providing climate knowledge support for society's decisionmakers will require developing an extremely accurate climate prediction capability to understand and accurately simulate Earth's complex energy-chemistry-climate system.

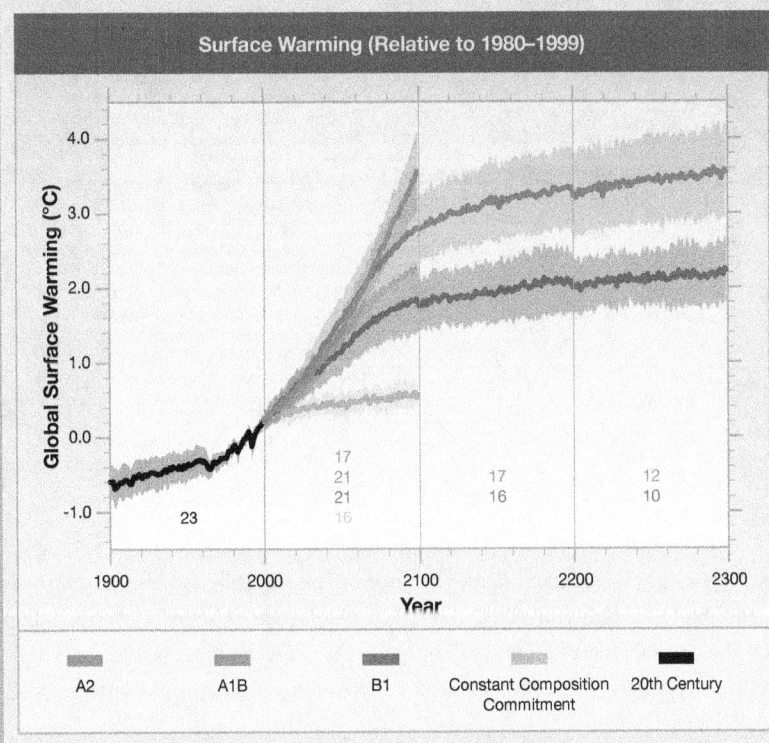

Surface Warming (Relative to 1980–1999)

Figure 1: Surface Warming (Relative to 1980-1999). Surface warming (relative to 1980–1999) from the WCRP multi-model data set for the scenarios of high, medium, and low emissions (A2, A1B, and B1, respectively), shown as continuations of the 20th century simulations that included combinations of natural and anthropogenic forcings. Values beyond 2100 are for idealized stabilization where greenhouse gas concentrations were held fixed at year 2100 values and the models were run to the year 2300 to assess climate change commitment. Similarly, the "constant composition commitment" experiment is for idealized stabilization of all greenhouse gas concentrations at year 2000 values, with the model calculations projected through the year 2100. Lines show the multi-model means; shading denotes the ±1 standard deviation range of individual model annual means. Discontinuities between different periods have no physical meaning and are caused by the fact that the number of models that have run a given scenario is different for each period and scenario, as indicated by the colored numbers given for each period and scenario at the bottom of the panel. *Credit: G.A. Meehl, NCAR (reproduced from the* **Bulletin of the American Meteorological Society** *with permission from the American Meteorological Society).*

11

EXAMPLES OF CCSP CONTRIBUTIONS
TO THE IPCC FOURTH ASSESSMENT REPORT [CONTINUED]

Figure 1 on the previous page summarizes the results from the IPCC Fourth Assessment Report experiments in terms of time series of globally averaged surface air temperatures from the different models for the various experiments. "Committed warming" is the term used to describe the lags between increases in concentrations of greenhouse gases and observed atmospheric temperature changes, caused by a number of factors including the thermal inertia of ocean and ice. Committed warming averages 0.1°C per decade for the first 2 decades of the 21st century. Across all scenarios, the average warming is 0.2°C per decade for that time period (the recent observed trend over roughly the last 2 decades is about 0.2°C per decade). Figure 2 shows that regional surface temperature changes are roughly double the globally averaged value, with the largest increases seen at high latitudes.

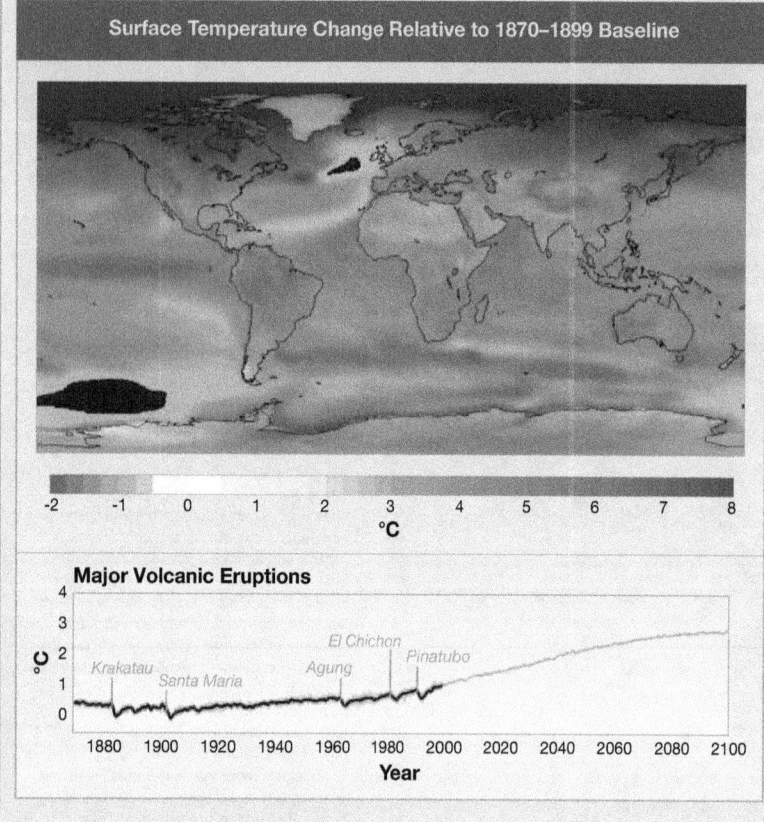

Surface Temperature Change Relative to 1870–1899 Baseline

Major Volcanic Eruptions

Krakatau   Santa Maria   Agung   El Chichon   Pinatubo

Figure 2: Surface Temperature Change Relative to 1870-1899 Baseline. The top panel shows that regional maximum surface temperature increases projected by the end of the 21st century are roughly double the globally averaged value, with the largest increases seen at high latitudes. Surface temperature increases in the tropics are much less, due to the more significant amounts of atmospheric water vapor, a very effective greenhouse gas. The lower panel shows instances of dramatic global cooling resulting from large volcanic eruptions. Such eruptions—for example, Krakatau in 1883— can be seen a number of times in this run as sharp, 5-year cooling events. As there is no method for predicting the occurrence of volcanic eruptions in the future, the smooth temperature curve from 2000 to 2100 reflects the fact that such events are not included in the climate models after 2000.
Credit: G. Strand, NCAR (adapted from the **Journal of Climate** with permission from the American Meteorological Society).

decisionmakers to use when planning for the future. The scientific assessment addresses all areas required in Section 106 of the Global Change Research Act of 1990, including impacts on the natural environment, agriculture, energy production and use, land and water resources, transportation, human health and welfare, human social systems, and biological diversity.

Examples of recent progress toward the five CCSP goals are briefly summarized below. Some of these examples were included in the Scientific Assessment.

Goal 1. Improve knowledge of the Earth's past and present climate and environment, including its natural variability, and improve understanding of the causes of observed variability and change.

Over the past 30 years, a combination of ground and global satellite observations together with Earth system models conducted by the CCSP-participating agencies have resulted in remarkable progress in first documenting and subsequently understanding Earth system components of the global climate system and how these components interact. These multi-agency activities have resulted in basic information that has led to the formulation of numerical simulation models required to further understanding of Earth's climate system. Numerical climate simulation models using observed radiative forcing provide the means to anticipate future climate, manage future climate risks, and explore opportunities related to climate variability and change. Model simulations also identify shortcomings in the climate observations, thereby providing the feedback needed to improve the cycle of measurement and modeling.

In 2006, it was reported that the ocean cooled from 2003 to 2005.[6] This conclusion was inconsistent with a wide variety of numerical modeling results and direct measurements of surface and lower atmosphere warming. As a consequence of this reported cooling, some suggested that global warming was not occurring. This situation was resolved through a reanalysis of the ocean temperature data from the Argo float array: a serious error in these data was found to be due to a combination of measurement and analysis artifacts. The authors of the original study subsequently published a correction to their original work.[7]

Publication of the correction of the ocean temperature trend from 2003 to 2005 was subsequent to CCSP Synthesis and Assessment Product 1.1, *Temperature Trends in the Lower Atmosphere*, published in 2006.[8] This report corrected errors found in earlier analyses of atmospheric data from satellites and weather balloons. Corrections to the satellite data analyses and corrections to the weather balloon measurements showed globally averaged, mid-troposphere temperatures increased by 0.1 to 0.2°C per decade from the late 1970s to the present.[9,10] These corrections resulted in mid-troposphere temperature trends similar to measured surface air temperatures for the same time period. This synthesis and assessment product increased confidence in understanding of the causes of climate change.

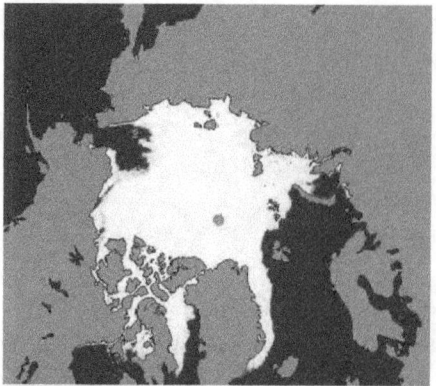

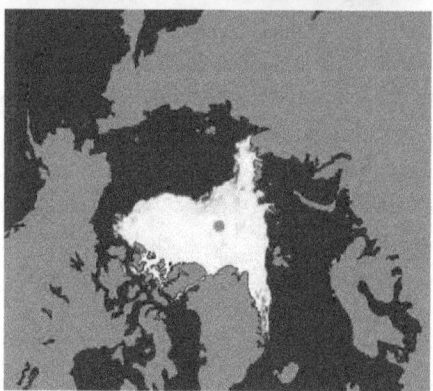

CCSP agencies have been very active in observing the Arctic, an area where numerical climate models project the greatest increase in surface temperatures. Because of the direct consequences of warmer surface temperatures for sea ice, permafrost, and glacier recession in the Arctic and their feedbacks to climate, a wide range of studies have been and are being performed there. For example, a combination of laser altimetry, passive microwave, and synthetic aperture radar satellite data has documented the unprecedented late-summer decline of Arctic sea ice extent in 2007 to a record minimum 23% lower than the previous record minimum observed in September 2005.[11] This work is complemented by several other studies from the Arctic region. For example, gravity data from the Gravity Recovery and Climate Experiment (GRACE) satellite show that Greenland is losing ice mass at an increasing rate while documenting the month-to-month ice mass variations. In 2006, Greenland's ice mass suffered a net annual loss of approximately 110 km^3 yr^{-1} due to losses exceeding gains as stratified by altitude. At altitudes above 2000 m, Greenland's ice mass increased at a rate of 50 km^3 yr^{-1}, consistent with climate forcing producing a more active hydrological cycle (i.e., increased snowfall). For altitudes below 2000 m, the loss due to melting was found to be 160 km^3 yr^{-1}.[12]

The upcoming CCSP Synthesis and Assessment Product 1.2, *Past Climate Variability and Change in the Arctic and at High Latitudes*, provides an assessment of progress toward Goal 1, summarizing research on climate change in the Arctic. High latitudes are especially sensitive to global warming and provide early indications of climate change. Furthermore, new paleoclimate data from the Arctic will provide a long-term context for recent observed temperature increases there.

CCSP Synthesis and Assessment Product 1.3, *Reanalyses of Historical Climate Data for Key Atmospheric Features: Implications for Attribution of Causes of Observed Change*, addresses understanding the magnitude of past climate variations. This is key to increasing present confidence in how and why climate has changed and why it may change in the future.

Goal 2. Improve quantification of the forces bringing about changes in the Earth's climate and related systems.

To understand the Earth's coupled ocean-atmosphere-land climate system, it is necessary to understand the abiotic and biotic processes that drive this system. These causes of climate change are called "forcing" factors and include greenhouse gases, land-cover change, volcanoes, air pollution and aerosols, and solar variability.

A 2007 report concluded that variations in the Sun are not responsible for recent global warming, either directly via increases in total solar irradiance or its ultraviolet spectral component, or through a decrease in the interaction between the solar wind and galactic cosmic rays.[13] This report is consistent with a significant body of literature on this topic, although it had been suggested that interaction between the variable solar wind and galactic cosmic rays influences climate forcing through variable cloud formation by some unknown mechanism.[14] Potential climate forcing by solar variability has been used to diminish human activities as a cause of global warming. While the authors of this report are European, they relied heavily upon NASA satellite data in their work.

CCSP agencies are making tremendous progress in understanding the role of aerosols in climate forcing. Aerosols are fine particles in the atmosphere that result from pollution, smoke, and dust. If they absorb light, they warm the atmosphere; if they reflect light, they cool it. The net forcing of aerosols is a key global warming uncertainty.

New Arctic research, much of it funded by CCSP agencies, has increased understanding of global climate change involving aerosols. A recent analysis of black carbon and non-sea salt sulfate aerosols from Greenland ice cores found a maximum black carbon net forcing of about 3 Wm^{-2} in the Arctic.[15] This maximum forcing occurred from 1906 to 1910 and was about eight times the pre-industrial forcing. Since the 1906 to 1910 period, the winter deposition of black carbon has gradually decreased while the summer deposition is highly variable, due to the variability of forest fires in North America. Non-sea salt sulfate aerosols, the product of fossil fuel combustion and a reflecting or "cooling" aerosol, increased strongly from the 1930s into the 1970s, after which they slowly declined until around 1992, and then plunged to pre-industrial levels by the early 2000s. Implementation of the Clean Air Act in the 1970s lowered U.S. sulfur emissions. Other countries followed with similar actions, explaining the dramatic reduction in non-sea salt sulfate aerosols since the 1970s.

CCSP work has shown how atmospheric composition, namely the gases and fine particles (aerosols) in the air, influence climate, the stratospheric ozone layer, and air quality. In the aerosol work, research has helped to better define the climate effects of

15

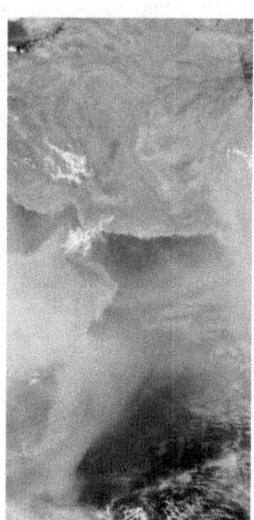

aerosols in large pollution plumes above regions of southern Asia (the "Asian brown cloud") using unmanned aerial vehicles, and in a variety of environments in the Arabian peninsula using a network of automated surface instruments.[16,17] Insights were also gathered regarding "Arctic haze," a widespread tropospheric haze or pollution cloud in the Arctic springtime. It was found that the Arctic haze resulted in an additional 2 to $3\ Wm^{-2}$ of atmospheric warming during the haze maximum in mid-April.[18] The climate effects of this phenomenon will be further studied in 2008 using aircraft and ship platforms, as well as ground-stations. The connections between aerosols, clouds, and climate—one of the areas of largest uncertainty in current understanding of climate change—were the subject of CCSP research in field and modeling studies. Data from a field study showed that the chemical composition of the aerosol particles (specifically, the organic content) influences how effectively the aerosol activates the formation of cloud droplets.[19] Other research showed that the aerosol chemical composition affects the very structure of clouds, in a way that has dramatic implications for how the clouds reflect light and hence affect climate.[20] The knowledge gained through these CCSP studies will lead to advances in the ability to accurately represent aerosol-cloud-climate processes in models, ultimately contributing to the development of an improved predictive capability.

Researchers in CCSP have shown that atmospheric constituents other than carbon dioxide (CO_2), including water vapor and ozone, have implications for Earth's climate. Pollution was shown to interact with the nitrogen oxides formed during lightning strikes to cause increases in the summertime amounts of ozone (a greenhouse gas) in the upper troposphere above eastern North America during summer.[21] An 8-year record of surface measurements has shown how water vapor affects climate-related cloud properties.[22] The climate implications of cloud-water vapor-aerosol interactions were the focus of an intense field study in the tropical tropopause region near Costa Rica, a highly active region of the atmosphere that affects the transport of gases from the lower atmosphere to the stratosphere. Satellite observations of carbon monoxide (CO), a trace gas related to pollution, have elucidated the processes that transport pollution from source regions to areas where air quality is impacted.[23] Other work is looking at the effects of climate change on regional air quality in the United States.[24] CCSP researchers have developed an improved formulation for gauging the anticipated future effect of ozone-depleting substances (many of which are themselves greenhouse gases) on the recovery of the stratospheric ozone layer.[25]

Analysis of observations from aircraft, towers, and radar shows that land-use patterns have a strong influence on the horizontal distribution of heat exchange between the surface and atmosphere. Combined with land surface model runs, the data suggest that soil moisture influences the relative magnitude of the horizontal variation of this heat

exchange.[26] This work provides additional support to the observation that a change in land cover from natural vegetation to managed uses not only affects the carbon cycle of the area in question, but also results in warmer surface temperatures.

CCSP research has investigated various carbon sequestration or carbon sink possibilities. On land, temperate and boreal forests in the Northern Hemisphere act as a substantial carbon sink, storing about 0.6 to 0.7 GtC per year. However, new results from the AmeriFlux research network show that forest disturbance from harvest and fire are responsible for much of the overall variability in forest carbon sequestration.[27] Forests are a carbon source to the atmosphere for as many as 20 years after these disturbance events, followed by a long period of carbon sequestration. After accounting for age and disturbance effects (e.g., wildfires, harvesting, infestations, etc.), low continuous levels of nitrogen deposition, largely the result of anthropogenic activities, appear to overwhelmingly account for additional carbon sequestration by these forests. This work is significant because it identifies the role of low-level nitrogen fertilization of Northern Hemisphere forests in the sequestration of carbon by terrestrial ecosystems.

CCSP has produced two synthesis and assessment products and has two more in preparation that summarize progress towards meeting Goal 2 (see page 29 for the complete list of Goal 2 products). The most recently completed, *The North American Carbon Budget and Implications for the Global Carbon Cycle* (2.2), summarizes sources and sinks of carbon in North America.[28] Among the findings of this report is the expectation that the net release of carbon to the atmosphere will increase as North America's emissions continue to increase and its terrestrial carbon sinks, primarily in re-growing forests, decline. The report states that actions to address this imbalance in North America's carbon budget will likely require a mix of options that includes emissions

reductions as well as sink enhancements. *Aerosol Properties and Their Impacts on Climate* (2.3) will address uncertainties about how different types of aerosols, both warming and cooling, may affect climate and thus how climate change might be affected by reductions in their emission rates. *Trends in Emissions of Ozone-Depleting Substances, Ozone Layer Recovery, and Implications for Ultraviolet Radiation Exposure* (2.4) will provide an update on trends in stratospheric ozone, ozone depleting gases, and ultraviolet radiation exposure; progress in improving model evaluations of the sensitivity of the ozone layer to changes in atmospheric composition and climate; and relevant implications for the United States.

Goal 3. Reduce uncertainty in the projections of how the Earth's climate and related systems may change in the future.

CCSP research has led to significant improvements in the ability to produce estimates of future Earth climates on time scales of years to centuries, using numerical simulation models initialized with measured radiative forcing obtained from ground and satellite observations.

Key findings include making significant progress in understanding the response time of modern-day ice sheets to external temperature forcing. Previously, it was thought that the massive Greenland and Antarctic Ice Sheets took thousands of years to respond to increasing external temperatures. Recent work has measured major dynamical ice sheet changes likely related to warmer temperatures on a time scale of years.[29] This work stresses the importance of (1) improving large-scale models of ice sheet dynamics and their response to warmer temperatures, because current numerical models do not presently simulate the observed ice sheet dynamics; and (2) continuing the suite of satellite climate observations of ice sheets, such as laser altimetry, gravity field measurements, Landsat, Moderate Resolution Imaging Spectrometer (MODIS), passive microwave, and synthetic aperture radar. This ice sheet work has enormous implications for projecting sea-level rise with increasing surface temperature: Greenland contains a volume of ice that is the equivalent of about 7 m of global sea-level rise and Antarctica's ice sheet is the equivalent of about 57 m of global sea-level rise.

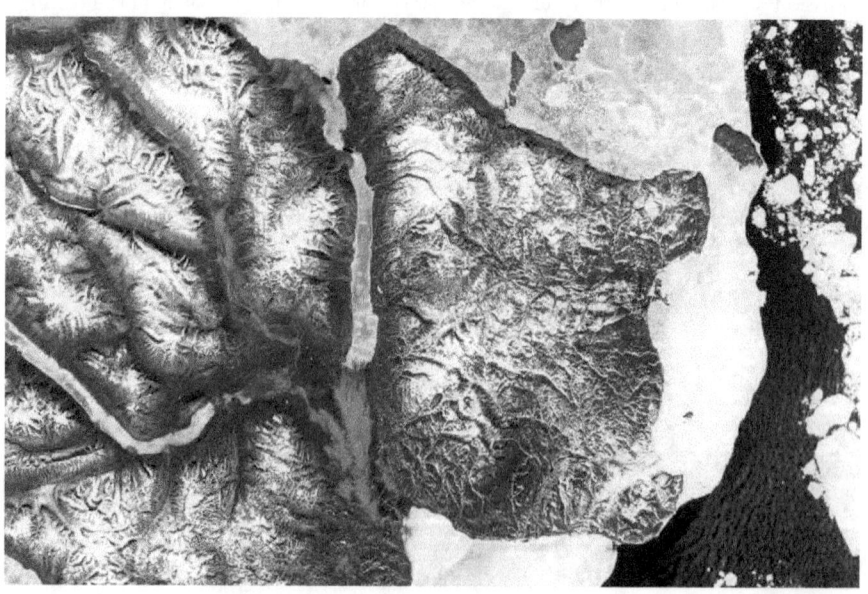

CCSP work has compared observations of Arctic sea ice with simulations of sea ice from more than a dozen models used in the IPCC Fourth Assessment Report. All models failed to produce the observed decreases in Arctic sea ice coverage, often substantially underestimating the decreases. This suggests that the Arctic could be seasonally free of sea ice earlier than IPCC projections.[11] A companion study considered factors that may contribute to declines in sea ice during the Arctic winter. The results show important regional differences in the processes leading to the changes, although the hemispheric-mean decline in winter ice extent appears primarily due to increasing sea surface temperatures in the Barents Sea and adjoining waters. These results emphasize the important roles of both atmospheric and oceanic processes in understanding and modeling the causes of rapid sea ice declines and demonstrate again that systematic climate observations are needed to understand the Earth's coupled land-atmosphere-ocean system.[10]

Important work on the relative contributions to sea-level rise from ice on land found that 60% of the sea-level rise was from ice loss from glaciers and ice caps and 40% was from the large ice sheets of Greenland and Antarctica.[11] All of these sources contribute less than the contribution to sea-level rise from the expansion of the oceans due to their warming.[1] If the loss of existing glacier ice from all sources occurs in proportion to projected increases in global temperature, sea level would rise about 0.5 to 1.4 m by 2100, which is greater than the estimate provided by IPCC (~0.2 to 0.6 m by 2100).[12] These studies stress the importance of climate change to sea-level rise through glacier recession and underscore many problems of human adaptation to sea-level rise.

CCSP Synthesis and Assessment Product 3.1, *Climate Models: An Assessment of Strengths and Limitations*, and Product 3.2, *Climate Projections Based on Emissions Scenarios for Long-Lived Radiatively Active Trace Gases and Future Climate Impacts of Short-Lived Radiatively Active Gases and Aerosols*, will constitute important steps toward reducing uncertainty in numerical model simulation of future climate conditions.

Goal 4. Understand the sensitivity and adaptability of different natural and managed ecosystems and human systems to climate and related global change.

A key purpose of CCSP is to understand potential effects of changes in the Earth's climate system on natural ecosystems, managed ecosystems, and human systems through a coordinated program of ecological observations, experimental research, and numerical simulations made possible by advanced observations and experimentation being planned and implemented by CCSP. Not only does this goal address water, air quality, health, human infrastructure, and agriculture, it also addresses the effects of climate change on natural terrestrial and oceanic ecosystems. This integration is being compiled in a series of seven CCSP synthesis and assessment products.

Studies from the Arctic document significant temperature increases in the upper layers of permafrost (soil that is permanently frozen).[13,14] The circum-Arctic permafrost stability is now being studied, with special attention to the potential for large fluxes of CO_2 and methane following thawing of these areas. This is an area where CCSP has and continues to stress the importance of ongoing and sustained study.

CCSP climate models projections for the 21st century for southwestern North America suggest much drier conditions, similar to the Dust Bowl of the 1930s.[15] This work indicates that storm tracks will shift northward, making the southwestern United States much drier and causing significant impacts on water resources, an increased frequency of forest fires, and potential economic destabilization. Snowmelt changes in mountains of the western United States have important implications on water availability in the southwestern regions.

CCSP-funded work has found that the ocean's ability to remove more CO_2 from the atmosphere will be impaired with warmer temperatures. The absorption of anthropogenic CO_2 and deposition of acid rain arising from fossil fuel and agricultural emissions can both contribute to the acidification of the ocean, reducing inorganic carbon storage. Additional impacts arise from atmospheric nitrogen deposition, leading to elevated primary production and biological drawdown of dissolved inorganic carbon that in some places reverses the sign of the surface acidity and air-sea CO_2 exchange. On a global scale, the alterations in surface water chemistry from anthropogenic deposition of reactive sulfur and nitrogen compounds are a few percent of the acidification, although the impacts are more substantial in coastal waters, where the ecosystem responses to ocean acidification could have the most severe implications for humanity.[16,17]

There has been substantial interagency cooperation over the past year in strengthening the capacity of the science community to incorporate an improved understanding of impacts and adaptation strategies in integrated assessments and in integrated modeling of the combined human and natural system behaviors, both in driving climate change and responding to climate change. During the past year, a major week-long session was held in Snowmass, Colorado, which brought together ecologists, Earth system modelers, economists, energy experts, health experts, social scientists, specialists in adaptation, and multiple agency representatives, all of whom had been important contributors to the recently published IPCC reports. The exchange of perspectives, review of IPCC results, and presentation of more recent research results have resulted in new collaborations being formed to investigate such issues as a more sophisticated integration

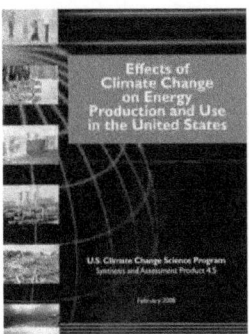

of Earth system models; impact, adaptation, and vulnerability models; and integrated assessment models and an exploration of how the different research communities approach problems such as land-use and land-cover change, with the intent of developing more coordinated approaches.

Synthesis and Assessment Product 4.5, *Effects of Climate Change on Energy Production and Use in the United States*, was published in October 2007.[18] The report summarizes the current knowledge base concerning the possible effects of global change on energy production and use in the United States. The authors surveyed and assessed the available literature, paying attention to research findings on the implications of climate variability for energy production and use, and identified and considered relevant studies carried out in connection with CCSP, CCTP, and other programs of CCSP agencies. Significant attention was directed to consulting relevant stakeholders such as the electric utility and energy industries, environmental nongovernmental organizations, and the academic research community to determine what analyses had been conducted and what reports issued. Besides addressing questions of possible direct effects of climate change on energy consumption and production in the United States, the report also considered how climate change might affect various factors that indirectly shape energy production and consumption, such as energy technology choices, energy institutional structures, regional economic growth, energy prices, energy security, and greenhouse gas emissions.

The research evidence is relatively clear that climate warming will mean reductions in total U.S. heating requirements for buildings and increases in total cooling requirements. These changes will vary by region and by season, but they will affect household and business energy costs and their demands on energy supply institutions. In general, the changes imply increased demand for electricity, which supplies virtually all cooling energy services but only some heating services. Secondly, the effects of climate change on energy production are less strong than on energy consumption, but climate change could affect energy production and supply (a) if extreme weather events become more intense; (b) where regions dependent on water supplies for hydropower and/or thermal power plant cooling face reductions in water supplies; (c) where temperature increases

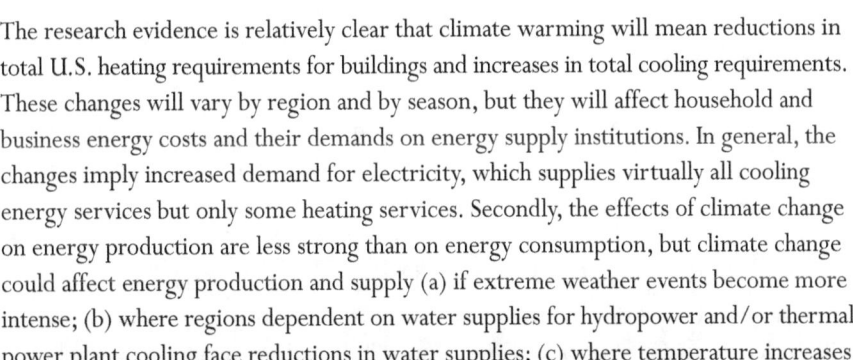

decrease overall thermoelectric power generation efficiencies; and (d) where changed conditions affect facility site decisions. Most effects are likely to be modest except for possible regional effects of extreme weather events and water shortages. Thirdly, results for indirect effects ranged from abundant information about possible effects of climate change policies on energy technology choices to extremely limited information about such issues as effects on energy security. Based on this mixed evidence, it appears that climate change is likely to affect risk management in the investment behavior of some energy institutions, and it is very likely to have some effects on energy technology research and development investments and energy resource and technology choices. In addition, climate change can be expected to affect other countries in ways that in turn affect energy conditions in the United States through its participation in global and hemispheric energy markets, and climate change concerns could interact with some driving forces behind policies focused on U.S. energy security.

Four synthesis and assessment products address how changing climate affects natural and managed ecosystems (i.e., 4.1, 4.2, 4.4, and 4.6). See the table on page 29 for a complete list of the Goal 4 products.

Goal 5. Explore the uses and identify the limits of evolving knowledge to manage risks and opportunities related to climate variability and change.

A critical responsibility for CCSP is to encourage the CCSP agencies to use scientific knowledge of climate and use of numerical simulation models to manage risks and identify opportunities related to climate variability and change.

One example of progress toward CCSP Goal 5 is a recent study of the water resource consequences of increasing tropical temperatures in the South American Andes, which will accelerate melting of tropical glaciers. This will affect the generation of hydroelectric power through diminished dry season river discharge, and directly affect the freshwater supply in many Andean communities.[19]

CCSP is generating three synthesis and assessment products under the Goal 5 rubric. Synthesis and Assessment Product 5.1, *Uses and Limitations of Observations, Data, Forecasts, and Other Projections in Decision Support for Selected Sectors and Regions*, finds that there is a great need for producing regional climate information. Not only will this enable further evaluation of the reliability of current information, it is also crucial to developing new applications of climate data to aid in managing risks and opportunities. Synthesis and

Assessment Product 5.2, *Best-Practice Approaches for Characterizing, Communicating, and Incorporating Scientific Uncertainty in Decisionmaking*, shows that improvements in how scientific uncertainty is evaluated and communicated are needed to reduce misunderstanding and misuse of this information. Synthesis and Assessment Product 5.3, *Decision Support Experiments and Evaluations Using Seasonal to Interannual Forecasts and Observational Data*, finds that climate variability is an important factor in resource planning and management at all levels of government, from the national level to the local level. Improved application of forecasts and data will benefit society in addressing climate change at all levels.

Other CCSP synthesis and assessment products are also of direct relevance to Goal 5. CCSP Synthesis and Assessment Product 2.1 consisted of two reports: *Scenarios of Greenhouse Gas Emissions and Atmospheric Concentrations* (Part A) and *Global-Change Scenarios: Their Development and Use* (Part B).[40] Part A used several integrated assessment models as the foundation for a small group of new global emissions scenarios leading to long-term stabilization of greenhouse gas concentrations. One of the most important implications of the work reported in Part A was the primacy of technology in addressing climate change—not only in the near term, but also in the long term, where investments in basic science and technology can lay the foundations for deployment of dramatically improved technologies. Part B reviewed and evaluated how the science and stakeholder communities define, develop, implement, and communicate scenarios in the global climate change context, and how this process might be enhanced or improved. This included a review of past scenario development and application efforts. The report applies three integrated assessment climate models in a comparison of five different scenarios of greenhouse gas emissions under alternative assumptions regarding long-term global climate goals. The CCSP report is the first to use several alternative models to evaluate multiple stabilization scenarios in this way.

Pilot studies in the Gulf Coast region and the Chesapeake Bay were undertaken to test different approaches to assessing the flow and use of climate change science information in decisionmaking, the factors and institutions that affect its use, and the types and characteristics of decisions most sensitive to climate change and most in need of additional reevaluation and research in light of projected changes. Results from these studies are being used to determine the applicability of a decision assessment approach to the national level and to decisions related to water quality.

The Climate Impacts Group, a CCSP-supported Regional Integrated Sciences and Assessments (RISA) team, and King County, Washington's Climate Team created a guidebook on preparing for and adapting to climate change. *Preparing for Climate Change: A Guidebook for Local, Regional, and State Governments* recommends a detailed, easy-to-understand process for climate change preparedness based on familiar resources and tools. Local Governments for Sustainability contributed to the production and dissemination of the guidebook to make it accessible to local governments across the United States. The results of this guidebook will be used in future planning for sectoral work on urban issues.

CCSP FY 2009 KEY INTERAGENCY IMPLEMENTATION ACTIVITIES

The program's long-term vision, mission, goals, and objectives are described in the CCSP Strategic Plan. Implementation of this long-term plan occurs through agency activities that often benefit significantly from ongoing CCSP-facilitated coordination. CCSP has identified several key areas for FY 2009 that require particularly strong interagency coordination to achieve success; they cannot be adequately addressed by one agency alone. Although these priorities are only a small part of the overall program, they are vital mechanisms through which CCSP will continue to integrate agency activities to create knowledge and products that are greater than the sum of the individual agency inputs. The development of CCSP interagency priorities is the result of a variety of planning processes, including planning processes within the 13 CCSP agencies (see Appendix A) and interagency planning conducted by the CCSP Interagency Committee (i.e., the Subcommittee on Global Change Research) and its subsidiary Interagency Working Groups. CCSP's interagency planning is informed by external advice from several NRC committees. CCSP's annual implementation priorities are logical evolutions of the program's interagency approaches to the priorities established in the CCSP Strategic Plan. The selection criteria for these activities require that they are founded upon a solid intellectual basis and are of high scientific quality; require coordination and/or integration across multiple CCSP agencies to create value-added products and services that cannot be created by any one agency alone; improve the characterization of key areas of scientific uncertainty and/or improve decision-support tools; provide a timely response to a particular need or leveraging opportunity; and are cost-effective.

The interagency implementation priorities generally represent only a fraction of CCSP's portfolio. The focus areas are listed here in an order similar to the research elements described in the CCSP Strategic Plan. However, due to their integrative nature they do not follow a one-to-one mapping to the research elements.

Quantification of Climate Forcing and Feedbacks by Aerosols, Non-CO$_2$ Greenhouse Gases, Water Vapor, and Clouds. This research will continue to use an array of observational approaches (satellite, intensive field campaigns using aircraft, ships, and balloons, and ground-based *in situ* and remote sensing networks) to quantify the abundances of atmospheric climate forcing agents and their precursors. The observations are being coupled with modeling analyses and laboratory studies to quantify the interrelated chemical and physical processes that control the magnitude and distribution of aerosols, clouds, tropospheric ozone, non-CO$_2$ trace gases, and water vapor. Developing an integrated understanding of the chemical and physical processes that determine the distribution and properties of these atmospheric constituents is a primary research emphasis, in order to improve the spatial and temporal representation of these atmospheric constituents in models and to improve the ability of those models to simulate current and future climate.

Assessing Abrupt Change in a Warming Climate: Examining the Feasibility of Developing an Abrupt Change Early Warning System. Paleoclimate research provides abundant evidence that major shifts in regional and global climate have occurred in the past over periods as short as a few years to decades. Previous work has focused on two primary areas: paleoclimate reconstructions to provide insight into past abrupt climate events, and mechanistic studies of potential causes for rapid changes. Paleoclimate studies have highlighted the importance of changes in the Atlantic meridional overturning circulation (sometimes called the thermohaline circulation) in producing past abrupt climate changes. Climate modeling studies have suggested specific mechanisms for producing such changes, especially varying freshwater inputs into the North Atlantic. Rapid changes are now being observed in high-latitude regions, especially in summer Arctic sea ice, but the future implications of these changes are uncertain. Observational evidence indicates that the mass of large ice sheets in Greenland and Antarctica has decreased in recent years, contrary to projections described in the IPCC Third Assessment Report. The reasons for these decreases are not well understood, but if they continue to increase they will have important implications for future rates of sea-level rise.

While considerable progress has been made in these and other areas, significant gaps remain that limit ability to provide early warning assessments of the likelihood of future abrupt climate change, either globally or over the United States, within this century. Addressing these gaps will require a focused national effort that includes collaboration across multiple agencies. Key elements include (1) an integrated Earth system observational analysis capability that will serve as a primary means for detecting and monitoring rapid changes in the Earth system, and that will provide the observational foundation for developing an abrupt change early warning system; (2) an improved decadal forecasting capability that blends predictions of natural variability together with scenarios of future radiative forcing to improve estimates of the likelihood of near-term rapid changes; (3) a vigorous paleoclimate research program to reconstruct and analyze causes of past abrupt climate changes; and (4) accelerated efforts to determine implications for society and ecosystems of abrupt change.

Development of an Integrated Earth System Analysis Capability: A Focus On Creating a High-Quality Record of the State of the Atmosphere and Ocean since 1979. A national capacity for Integrated Earth System Analysis (IESA) is being developed that extends beyond current attempts to map individual components of the Earth system separately. Such a capacity is a fundamental prerequisite to understanding the coupling of, and feedbacks within, the Earth system (i.e., the atmosphere, ocean, land surface, and cryosphere, including their physical, chemical, and biological components). It is also fundamental to advancing climate prediction capabilities, whether on seasonal or multi-year to decadal time scales, because these predictions intrinsically involve coupling of Earth system components. Moreover, an IESA capability is required to realize the full value of measurements made by GEOSS.

Development of an End-to-End Hydrologic Projection and Application Capability. The development of an end-to-end water cycle/hydrologic projection and application capability will require improvements in (1) the characterization and parametric representation of water cycle processes in Earth system models; (2) the ability to downscale output from global Earth system models to improve the quality of the input used in hydrologic modeling at watershed and river basin scales; and (3) guidance on how to effectively use hydrologic model output to inform specific, long-term hydrology-related planning and policy decisions. This set of activities will examine the reliability and uncertainties in U.S. outcomes from different climate forcing scenarios, and will explore different approaches for making projections of surface and subsurface hydrologic conditions for the coming decades. Outputs will include improved parameterizations and projections of soil moisture, surface runoff and river flows, groundwater recharge, and base flow that characterize not only changes in the mean state, but also possible changes in patterns of variability. This activity will use a broad range of specific decisionmaking contexts as prototypes to explore approaches for improving the use of research outputs to inform decisionmaking related to long-term hydrologic changes. These decisionmaking contexts include not only water resource planning, but also other sectors that are affected by changes in hydrologic conditions. One specific example is the set of effects on public health, including the effects of changing hydrologic conditions on the abundance and distribution of disease vectors and on water quality.

Enhanced Carbon Cycle Research on High-Latitude Systems. There is evidence that warming is occurring in some high-latitude areas, and published research suggests that it may be affecting ecosystem function and structure. The IPCC Fourth Assessment Report stated that average Arctic temperatures increased by nearly twice the global average rate over the past 100 years, and that sea ice extent decreased on average by 2.7% per decade since 1978. Temperatures at the top of the permafrost layer have generally increased by up to 3°C since the 1980s. High-latitude temperature increases, predicted by most general circulation models, may influence carbon storage and release in both Arctic and sub-Arctic soils and ecosystems. Since these ecosystems contain nearly 40% of the global soil carbon that could potentially influence climate change, thawing associated with the predicted temperature increases could produce a substantial release of CO_2 and methane to the atmosphere and oceans, further perturbing climate. Although it is highly uncertain at what rate these gases might be released, it has been suggested that the release could occur relatively rapidly. High-latitude warming also affects the ability of the polar oceans to absorb CO_2 and other greenhouse gases.

Ecological Forecasting. The CCSP ecosystems component addresses three high-level research goals: (1) improve understanding of the feedbacks between ecosystems and global change, especially climate change; (2) determine potential consequences of global change for ecosystems; and (3) provide options for sustaining and improving ecosystems and their provision of goods and services. This requires the development of ecological forecasting capabilities to test understanding of these feedbacks, making

projections of the potential consequences, and assessing the ability of various policy options and decisions to sustain and improve ecosystems of concern. Ecological forecasting will require advances in experimental ecological research, advances in ecological observing networks, and scientific synthesis across the

physical, biological, and social sciences. Successful ecological forecasting will be related to interdisciplinary work across the CCSP working groups in order to understand and project the potential effects of human interactions with natural systems and the feedbacks from these interactions. There will no doubt be limits to what can be forecast, but discovering these limits and their causes will only enhance overall understanding of the ecosystems to manage and preserve. Likewise, explicit statements regarding uncertainties and estimates of error associated with the forecasts are essential. An FY 2009 priority for enhancing the ability to generate reliable ecological forecasts should take advantage of the full suite of research supported by the Ecosystems Interagency Working Group membership. Ecological forecasts will allow the incorporation of observations, experimental results, process studies, and modeling activities at a wide variety of scales ranging from molecular through regional and even global, addressing the needs of basic science researchers, and the agencies that support them. Development of these ecological forecasts also fulfills the requirements of natural resource managers and policymakers who seek to understand the effects of particular policies or management approaches on ecosystem function. Improved ecological forecasting capabilities will play an important role in informing decisionmakers about the potential effects of climate change and in supporting the development of policy instruments and management actions to address anticipated ecosystem changes.

DECISION SUPPORT: INFORMATION TO SUPPORT POLICY DEVELOPMENT AND ADAPTIVE MANAGEMENT

CCSP sponsors and conducts research that is ultimately related to policy and adaptive management decisionmaking. CCSP's decision-support approach is guided by several general principles, including:

- Early and continuing involvement of stakeholders
- Explicit treatment of uncertainties
- Transparent public review of analysis questions, methods, and draft results
- Evaluation of lessons learned from ongoing and prior decision-support and assessment activities.

Synthesis and Assessment Products

As noted previously, CCSP is generating a suite of synthesis and assessment products that integrate research results focused on key issues and related questions frequently raised by decisionmakers. Current evaluations of the science can be used for informing public debate, policy development, and adaptive management decisions and for defining

SYNTHESIS AND ASSESSMENT PRODUCTS

CCSP Goal 1. Improve knowledge of the Earth's past and present climate and environment, including its natural variability, and improve understanding of the causes of observed variability and change.

1.1 Temperature Trends in the Lower Atmosphere: Steps for Understanding and Reconciling Differences

1.2 Past Climate Variability and Change in the Arctic and at High Latitudes

1.3 Re-Analyses of Historical Climate Data for Key Atmospheric Features: Implications for Attribution of Causes of Observed Change

CCSP Goal 2. Improve quantification of the forces bringing about changes in the Earth's climate and related systems.

2.1 Scenarios of Greenhouse Gas Emissions and Atmospheric Concentrations (Part A) and Global-Change Scenarios: Their Development and Use (Part B)

2.2 North American Carbon Budget and Implications for the Global Carbon Cycle

2.3 Aerosol Properties and Their Impacts on Climate

2.4 Trends in Emissions of Ozone-Depleting Substances, Ozone Layer Recovery, and Implications for Ultraviolet Radiation Exposure

CCSP Goal 3. Reduce uncertainty in projections of how the Earth's climate and environmental systems may change in the future.

3.1 Climate Models: An Assessment of Strengths and Limitations

3.2 Climate Projections based on Emissions Scenarios for Long-Lived Radiatively Active Trace Gases and Future Climate Impacts of Short-Lived Radiatively Active Gases and Aerosols

3.3 Weather and Climate Extremes in a Changing Climate: Regions of Focus - North America, Hawaii, Caribbean, and U.S. Pacific Islands

3.4 Abrupt Climate Change

CCSP Goal 4. Understand the sensitivity and adaptability of different natural and managed ecosystems and human systems to climate and related global changes.

4.1 Coastal Elevations and Sensitivity to Sea-Level Rise

4.2 Thresholds of Change in Ecosystems

4.3 The Effects of Climate Change on Agriculture, Land Resources, Water Resources, and Biodiversity

4.4 Preliminary Review of Adaptation Options for Climate-Sensitive Ecosystems and Resources

4.5 Effects of Climate Change on Energy Production and Use in the United States

4.6 Analyses of the Effects of Global Change on Human Health and Welfare and Human Systems

4.7 Impacts of Climate Variability and Change on Transportation Systems and Infrastructure: Gulf Coast Study

CCSP Goal 5. Explore the uses and identify the limits of evolving knowledge to manage risks and opportunities related to climate variability and change.

5.1 Uses and Limitations of Observations, Data, Forecasts, and Other Projections in Decision Support for Selected Sectors and Regions

5.2 Best Practice Approaches for Characterizing, Communicating, and Incorporating Scientific Uncertainty in Decisionmaking

5.3 Decision Support Experiments and Evaluations using Seasonal to Interannual Forecasts and Observational Data

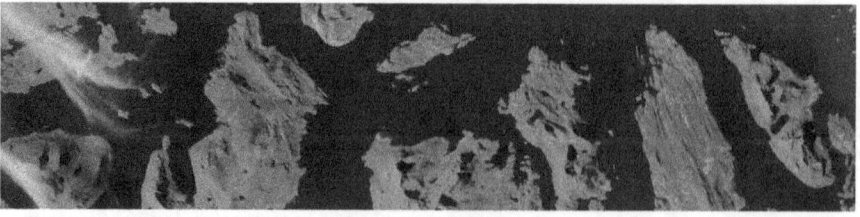

and setting the future direction and priorities of the program. The synthesis and assessment products constitute an important form of topic-driven integration of U.S. global change assessment efforts. These CCSP products are U.S. Government reports, subject to the provisions of the Information Quality Act (Section 515 of the Treasury and General Government Appropriations Act of 2001) and the Federal Advisory Committee Act Amendments of 1997 (PUB. L. 105-153, SEC. 2(A), (B), DEC. 17, 1997, 111 STAT. 2689). Findings from several of the synthesis and assessment products were included in CCSP's recently released *Scientific Assessment of the Impacts of Climate Change on the United States*.

The synthesis and assessment products are generated by researchers in a process that involves review by experts, public comment from stakeholders and the general public, and final approval by the departments/agencies involved in CCSP. Formal endorsement of the products by the Federal Government enhances their value for decisionmakers and the public at large. The program has prepared guidelines that describe steps to be followed in each of three phases of the preparation process: developing the prospectus, drafting and revising, and final approval and publication. This methodology for product development facilitates involvement of the research community and user groups in ensuring that the products are focused in a useful fashion and meet the highest standards of scientific excellence. The guidelines also encourage transparency by ensuring that public information about the status of the products will be provided through the *Federal Register*, on the CCSP web site, and by other means throughout the review and clearance process. If further clarification of specific issues is required, the NRC will provide advice on an as-needed basis to the lead agency responsible for the preparation of each product.

A list of the titles of the 21 synthesis and assessment products, arranged by CCSP goal, is given in the table on the preceding page.

As of June 2008, eight products have been released and several others are nearing completion. Up-to-date information on production status can be obtained from <www.climatescience.gov/Library/sap/sap-summary.php>, including details on opportunities for public comment on draft products.

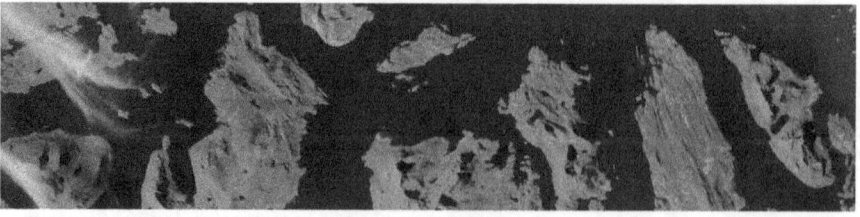

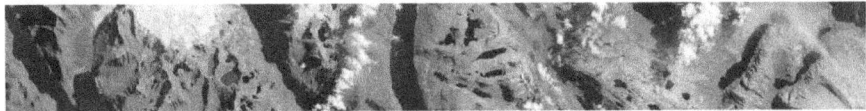

"Lessons Learned" in Decision Support and Assessment

To build on the experiences of earlier assessment activities, CCSP requested that the NRC carry out an analysis of global change assessments that have addressed topics broadly similar to those encompassed by CCSP. The study, released in early 2007, included a comparative analysis of past assessments that address issues directly related to the science and technical issues of CCSP. The committee concluded that global change assessments are critical for informing decisionmakers. In identifying essential properties of a successful assessment, it stressed that future assessment processes must communicate relevant information to the user, address the technical quality of the information, and demonstrate fairness and impartiality in the assessment process. The report identifies a number of essential elements that increase the probability that an assessment will effectively inform decisionmakers and other target audiences. CCSP is taking into account the findings of the NRC in its strategic planning efforts.

OUTLINE OF RESEARCH ELEMENT ACTIVITIES

The CCSP-participating agencies coordinate scientific research through a set of linked interdisciplinary research elements and cross-cutting activities that encompass a wide range of interconnected issues of climate and global change. Chapters 3 to15 of the CCSP Strategic Plan contain more detailed discussions of the research elements as well as activities that cut across all areas of the program. This report focuses on highlights of recent research and program plans for FY 2009.

Atmospheric Composition. The composition of the atmosphere at global and regional scales influences climate, air quality, stratospheric ozone, and precipitation, which in turn affect human health and the vitality of ecosystems. Research and observational activities coordinated and supported by CCSP are being used to assess how human activities and natural processes affect atmospheric composition, and how that understanding may be used to inform decisionmaking in the United States and abroad. In FY 2009, emphasis will be placed on studies of interactions between aerosols and non-CO_2 gases, enhanced measurements of atmospheric water vapor, and interactions of pollutants with climate change. Special emphasis will be placed on the climate impacts of pollutants associated with aviation.

See CCSP Strategic Plan Chapter 3.

31

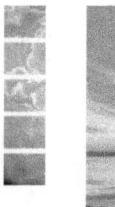

Climate Variability and Change (including Climate Modeling). Recognizing that the climate system operates seamlessly across a wide spectrum of time scales, CCSP-supported research encompasses both short-term climate variability and longer term climate change. Addressing the interaction of climate processes across time scales poses challenges not only in designing observation systems to monitor the climate system adequately, but also in constructing models that can properly reproduce its past behavior, and confidently project its future behavior. Earth system models, in combination with global Earth observations, must produce internally consistent maps of atmospheric, oceanic, land surface, and ice conditions both in near-real-time and retrospectively. These maps, or "analyses," will provide decisionmakers with tools to visualize the evolving state of the full climate system over the entire planet, and researchers with the ability to better explain observed changes in the climate system.

See CCSP Strategic Plan Chapters 4 and 10.

Global Water Cycle. Research associated with this element involves studies of the crucial role the water cycle plays in climate variability and change, and the influence climate has on aspects of the global water cycle on which society and nature depend. Through countless interactions within the Earth system, the global water cycle integrates physical, chemical, and biological processes that sustain ecosystems and influence climate and related global change. The ultimate goal of CCSP water cycle research is to provide a better foundation for decisions and investments by policymakers, managers, and individuals. Achieving this goal requires a program of activities that test predictions and data products in real decision contexts, demonstrate techniques and their effectiveness to potential users, and provide tools and strategies to transfer the science from the experimental realm to operations. In FY 2009, emphasis will be placed on coordinated observations and modeling of selected sites, at the river basin or catchment scale, to improve understanding of terrestrial water cycle processes leading to better closure constraints on water budgets at this scale. The improvements to land surface and hydrological models resulting from this research will lead to an enhanced ability to more accurately represent global change projections at the regional scales that affect water resources and other applications sectors dependent on the water cycle.

See CCSP Strategic Plan Chapter 5.

Land-Use and Land-Cover Change. Land use and land cover are linked to climate and weather in complex ways and are critical inputs for modeling greenhouse gas emissions, carbon balance, and ecosystems. Land-use and land-cover change (LULCC) studies have provided critical inputs to large-scale biomass and forest cover assessments; future LULCC goals include reducing uncertainties in biomass estimates, understanding regional heterogeneities in changes, and quantifying linkages and feedbacks between LULCC, climate change, and other human and environmental components. Research that examines historic, current, and future LULCC, its drivers, feedbacks to climate, and its environmental, social, economic, and human health consequences is therefore of utmost importance and often requires interagency and intergovernmental cooperation. Research plans focus on how management practices may change as climate and conservation policies change, and on feedbacks related to environmental, social, economic, and human health.

See CCSP Strategic Plan Chapter 6.

Global Carbon Cycle. Increasing levels of atmospheric CO_2 and methane are major drivers of climate change. The CCSP global carbon cycle element seeks to better quantify and understand the dynamics of the global carbon cycle that determine CO_2 and methane fluxes and carbon storage in terrestrial and oceanic ecosystems. Carbon cycle processes depend on climate, thus linking carbon cycle and climate change analyses is critical. Carbon cycle research involves multiple disciplines and extends over a broad range of spatial and temporal scales. Major multi-agency activities include the North American Carbon Program (NACP), an effort to describe and reduce uncertainties about the North American carbon budget and underlying processes, and the Ocean Carbon and Climate Change (OCCC) Program, a research effort aimed at determining how climate change will affect the future behavior of the oceanic carbon sink. In FY 2009, NACP will address key gaps and uncertainties in the carbon syntheses developed previously, and aspects of the OCCC and NACP will be coordinated to better quantify and understand the roles of adjacent ocean basins in the North American carbon budget. NASA will launch the Orbiting Carbon Observatory (OCO) to provide, for the first time, consistent atmospheric carbon observations globally from space, and carbon data assimilation systems will begin to derive estimates of carbon sources and sinks from these measurements.

See CCSP Strategic Plan Chapter 7.

Ecosystems. This research element studies the potential effects of global change on goods and services provided by aquatic and terrestrial ecosystems, using observations, experiments, modeling, and syntheses to focus on critical emerging questions. Newly initiated projects in terrestrial ecosystems are addressing cause-and-effect relationships between climatic variability and change and the distribution, abundance, and productivity of native and invasive organisms. Research is continuing into understanding how increasing CO_2 levels affect plants and microorganisms. Research in a Chesapeake Bay ecosystem is generating data to evaluate and forecast effects of warming, changes in fishing pressure, and eutrophication on economically important estuarine ecosystems. In the ocean, coral reef research is helping scientists and managers identify climatic and non-climatic stressors and thereby better manage these important ecosystems.

See CCSP Strategic Plan Chapter 8.

Decision-Support Resources Development and Related Research on Human Contributions and Responses. Decisionmakers and other interested citizens need reliable science-based information to make informed judgments regarding policy and actions to address the risks and opportunities of variability and change in climate and related systems. A wide variety of CCSP decision-support resources and related research on human contributions and responses is targeted at that objective. The outcomes of these activities are intended to inform public discussion of climate-related issues and scientifically assess and expand options for mitigation of and adaptation to climate variability and change. The most prominent of CCSP's ongoing decision-support activities is its synthesis and assessment process that involves preparation of 21 different products intended to support public discussion of climate science issues of particular importance to U.S. decisions. CCSP's research on human contributions and responses to global environmental variability and change includes analyses of human drivers of change and their potential impact, societal resilience and ways of reducing vulnerability, approaches for improving the ability of decisionmakers to utilize scientific information, and the effects of global environmental change on infrastructure and human health. CCSP's research is paying particular attention to aspects of global change that have greatest relevance to society, including drought and extreme or abrupt climate events.

See CCSP Strategic Plan Chapters 9 and 11.

Observing and Monitoring the Climate System. CCSP provides active stewardship of observations that document the evolving state of the climate system, that allow for improved understanding of its changes, and that contribute to an improved predictive capability for society. Some of these observations are not part of the CCSP budget (e.g., operational satellites such as NPOESS) but are crucial to its success. A core CCSP activity is U.S. participation in the broad-based strategy of the international Global Climate Observing System (GCOS) in monitoring atmospheric, oceanic, and terrestrial domains with an appropriate balance of *in situ* and remotely sensed observations. As the U.S. plan for climate observations moves forward, it strives to build on the *GCOS Implementation Plan* (see <www.wmo.int/pages/prog/gcos/Publications/gcos-92.pdf>). CCSP endorses the use of this plan as a blueprint for guiding GCOS-related climate observation activities documented in the chapter on "Observing and Monitoring the Climate System." In FY 2009, observing activities by CCSP agencies will continue to focus on monitoring the polar climate as part of the International Polar Year (IPY) series of international cooperative studies. IPY plans to advance polar observations by establishing a variety of new multidisciplinary observatories using the latest technologies in sensor web (network of spatially distributed sensor platforms that wirelessly communicate with each other) and power-efficient designs. Data from these, as well as more traditional surface- and space-based observatories, will initiate long-term, high-quality sustained measurements needed to detect future climate change. The United States plans to increase its efforts to observe the polar atmosphere, ice, and ocean, as well as to leverage its investments in polar research with international partners. A continuing challenge to CCSP agencies is ensuring the long-term integrity and understandability of data products provided by remote-sensing and *in situ* observing systems. Key parts of this challenge include continuing to integrate surface climate observations via the Climate Reference Network and modernized Historical Climatology Network; expanding the GCOS observing network via activities such as the Atmospheric Radiation Measurement Climate Research Facility and the GCOS Reference Upper Air Network; and support for a number of research-related satellite missions described in Chapter 8 of this report.

See CCSP Strategic Plan Chapters 12 and 13.

Communication. CCSP's member agencies support a broad array of communication initiatives. CCSP has developed a strategy and implementation plan for helping to coordinate and facilitate these activities. These efforts are intended to improve public understanding of climate change research by disseminating the results of CCSP activities credibly and effectively, and by making CCSP science findings and products available to a diverse set of audiences. CCSP facilitates communication of the results of individual agencies, as well as providing coordination in communicating the results of climate activities of the Federal Government.

See CCSP Strategic Plan Chapter 14.

International Research and Cooperation. CCSP, through its working groups including the Interagency Working Group on International Research and Cooperation, participates in and provides input to major international scientific and related organizations on behalf of the U.S. Government and scientific community. CCSP also provides support to maintain the central infrastructure of several international research programs and international activities that complement CCSP and U.S. Government goals in climate science.

See CCSP Strategic Plan Chapter 15.

THE U.S. CLIMATE CHANGE SCIENCE PROGRAM FOR FY 2009
CHAPTER REFERENCES

1) **IPCC**, 2007: Summary for Policymakers. In: *Climate Change 2007: The Physical Science Basis. Contribution of Working Group I to the Fourth Assessment Report of the Intergovernmental Panel on Climate Change* [Solomon, S., D. Qin, M. Manning, Z. Chen, M. Marquis, K.B. Averyt, M. Tignor, and H.L. Miller (eds.)]. Cambridge University Press, Cambridge, United Kingdom and New York, NY, USA.

2) **NSTC** Joint Subcommittee on Ocean Science and Technology, 2007: *Charting the Course for Ocean Science in the United States for the Next Decade: An Ocean Research Priorities Plan and Implementation Strategy.* Washington, DC, 84 pp. Available at <ocean.ceq.gov/about/docs/orppfinal.pdf>.

3) **Meehl**, G.A., C. Covey, T. Delworth, M. Latif, B. McAvaney, J.F.B. Mitchell, R.J. Stouffer, and K.E. Taylor, 2007: The WCRP CMIP3 multi-model dataset: A new era in climate change research. *Bulletin of the American Meteorological Society*, **88(9)**, 1383-1394, doi:10.1175/BAMS-88-9-1383.

4) **Meehl**, G.A., W.M. Washington, B.D. Santer, W.D. Collins, J.M. Arblaster, A. Hu, D. Lawrence, H. Teng, L.E. Buja, and W.G. Strand, 2006: Climate change projections for twenty-first century and climate change commitment in the CCSM3. *Journal of Climate*, **19**, 2597-2616.

5) **Thompson**, L.G., E. Mosley-Thompson, H. Brecher, M. Davis, B. León, D. Les, P.-N. Lin, T. Mashiotta, and K. Mountain, 2006: Abrupt tropical climate change: Past and present. *Proceedings of the National Academy of Sciences*, **103**, 10536-10543, doi:10.1073/pnas.0603900103.

6) **Lyman**, J.M., J.K. Willis, and G.C. Johnson, 2006: Recent cooling of the upper ocean. *Geophysical Research Letters*, **33(18)**, L18604, doi:10.1029/2006GL027033.

7) **Willis**, J.K., J.M. Lyman, G.C. Johnson, and J. Gilson, 2007: Correction to "Recent cooling of the upper ocean." *Geophysical Research Letters*, **34(16)**, L16601, doi:10.1029/2007GL030323.

8) **CCSP**, 2007: *Temperature Trends in the Lower Atmosphere: Steps for Understanding and Reconciling Differences.* A Report by the Climate Change Science Program and the Subcommittee on Global Change Research [Karl, T.R., S.J. Hassol, C.D. Miller, and W.L. Murray (eds.)]. Department of Commerce, National Oceanic and Atmospheric Administration, Washington, DC, 164 pp.

9) **Vinnikov**, K.Y., N.C. Grody, A. Robock, R.J. Stouffer, P.D. Jones, and M.D. Goldberg, 2006: Temperature trends at the surface and in the troposphere. *Journal of Geophysical Research*, **111**, D03106, doi:10.1029/2005jd006392.

10) **Sherwood**, S.C., J. Lanzante, and C. Meyer, 2005: Radiosonde daytime biases and late 20th century warming. *Science*, **309**, 1556-1559.

11) **Stroeve**, J., M. Serreze, S. Drobot, S. Gearheard, M. Holland, J. Maslanik, W. Miere, and T. Scambos, 2008: Arctic sea ice extent plummets in 2007. *EOS Transactions of the American Geophysical Union*, **89(2)**, 13-14.

12) **Luthcke**, S.B., H.J. Zwally, W. Abdalati, D.D. Rowlands, R.D. Ray, R.S Nerem, F.G. Lemoine, J.J. McCarthy, and D.S. Chinn, 2006: Recent Greenland ice mass loss by drainage system from satellite gravity observations. *Science*, **314**, 1286-1289, doi:10.1126/science.1130776.

13) **Lockwood**, M. and C. Frohlich, 2007: Recent oppositely directed trends in solar climate forcings and the global mean surface air temperature. *Proceedings of the Royal Society A*, **463**, 2447-2460, doi:10.1098/rspa.2007.1880.

14) **Svensmark**, H., 2007: Cosmoclimatology: a new theory emerges. *Astronomy and Geophysics*, **48(1)**, 18-24.

15) **McConnell**, J.R., R. Edwards, G.L. Kok, M.G. Flanner, C.S. Zender, E.S. Saltzman, J.R. Banta, D.R. Pasteris, M.M. Carter, and J.D.W. Kahl, 2007: 20th-century industrial black carbon emissions altered arctic climate forcing. *Science*, **317**, 1381-1384.

16) **Ramanathan**, V., M.V. Ramana, G. Roberts, D. Kim, C. Corrigan, C. Chung, and D. Winker, 2007: Warming trends in Asia amplified by brown cloud solar absorption. *Nature*, **448**, 575-578.

17) **Eck**, T.F., B.N. Holben, J.S. Reid, A. Sinyuk, O. Dubovik, A. Smirnov, D. Giles, N.T. O'Neill, S.-C. Tsay, Q. Ji, A. Al Mandoos, M. Ramzan Khan, E.A. Reid, J.S. Schafer, M. Sorokine, W. Newcomb, and I. Slutsker, 2008: Spatial and temporal variability of column-integrated aerosol optical properties in the southern Arabian Gulf and United Arab Emirates in summer. *Journal of Geophysical Research*, **113**, D01204, doi:10.1029/2007JD008944.

THE U.S. CLIMATE CHANGE SCIENCE PROGRAM FOR FY 2009
CHAPTER REFERENCES (CONTINUED)

18) **Quinn**, P.K., G. Shaw, E. Andrews, E.G. Dutton, T. Ruoho-Airola, and S.L. Gong, 2007: Arctic haze: current trends and knowledge gaps. *Tellus*, **59B(1)**, 99-114.

19) **Sorooshian**, A., M.-L. Lu, F.J. Brechtel, H. Fonsson, G. Feingold, R.C. Flagan, and J.H. Seinfeld, 2007: On the source of organic acid aerosol layers above clouds. *Environmental Science and Technology*, **41**, 4647-4654.

20) **Xue**, H., G. Feingold, and B. Stevens, 2008: Aerosol effects on clouds, precipitation, and the organization of shallow cumulus convection. *Journal of Atmospheric Science*, **65**, 392-406, doi:10.1175/2007JAS2428.1.

21) **Cooper**, O.R., M. Trainer, A.M. Thompson, S.J. Oltmans, D.W. Tarasick, J.C. Witte, A. Stohl, S. Eckhardt, J. Lelieveld, M.J. Newchurch, B.J. Johnson, R.W. Portmann, L. Kalnajs, M.K. Dubey, T. Leblanc, I.S. McDermid, G. Forbes, D. Wolfe, T. Carey-Smith, G.A. Morris, B. Lefer, B. Rappenglück, E. Joseph, F. Schmidlin, J. Meagher, F.C. Fehsenfeld, T.J. Keating, R.A. Van Curen, and K. Minschwaner, 2007: Evidence for a recurring eastern North America upper tropospheric ozone maximum during summer. *Journal of Geophysical Research*, **112**, D23304, doi:10.1029/2007JD008710.

22) **Dong**, X., B. Xi, and P. Minnis, 2006: Observational evidence of changes in water vapor, clouds, and radiation at the ARM SGP site. *Geophysical Research Letters*, **33**, L19818, doi:10.1029/2006GL027132.

23) **Jiang**, J.H., N.J. Livesey, H. Su, L. Neary, J.C. McConnell, and N.A.D. Richards, 2007: Connecting surface emissions, convective uplifting, and long-range transport of carbon monoxide in the upper troposphere: New observations from the Aura Microwave Limb Sounder. *Geophysical Research Letters*, **34**, L18812, doi:10.1029/2007GL030638.

24) **Tao**, Z., A. Williams, H.-C. Huang, M. Caughey, and X.-Z. Liang, 2007: Sensitivity of U.S. surface ozone to future emissions and climate changes. *Geophysical Research Letters*, **34**, L08811, doi:10.1029/2007GL029455.

25) **Newman**, P.A., J.S. Daniel, D.W. Waugh, and E.R. Nash, 2007. A new formulation of equivalent effective stratospheric chlorine (EESC). *Atmospheric Chemistry and Physics*, **7**, 4537-4552.

26) **LeMone**, M.A., F. Chen, J.G. Alfieri, M. Tewari, B. Geerts, Q. Miao, R.L. Grossman, and R.L. Coulter, 2007: Influence of land cover and soil moisture on the horizontal distribution of sensible and latent heat fluxes in southeast Kansas during IHOP_2002 and CASES-97. *Journal of Hydrometeorology*, **8(1)**, 68-87.

27) **Magnani**, F., M. Mencuccini, M. Borghetti, P. Berbigier, F. Berninger, S. Delzon, A. Grelle, P. Hari, P.G. Jarvis, P. Kolari, A.S. Kowalski, H. Lankreijer, B.E. Law, A. Lindroth, D. Loustau, G. Manca, J.B. Moncrieff, M. Rayment, V. Tedeschi, R. Valentini, and J. Grace, 2007: The human footprint in the carbon cycle of temperate and boreal forests. *Nature*, **447(7146)**, 849-851.

28) **CCSP**, 2007. *The First State of the Carbon Cycle Report (SOCCR): The North American Carbon Budget and Implications for the Global Carbon Cycle*. A Report by the U.S. Climate Change Science Program and the Subcommittee on Global Change Research [King, A.W., L. Dilling, G.P. Zimmerman, D.M. Fairman, R.A. Houghton, G. Marland, A.Z. Rose, and T.J. Wilbanks (eds.)]. National Oceanic and Atmospheric Administration, National Climatic Data Center, Asheville, NC, USA, 242 pp.

29) **Bamber**, J.L., R.B. Alley, and I. Joughin, 2007: Rapid response of modern day ice sheets to external forcing. *Earth and Planetary Science Letters*, **257(1-2)**, 1-13.

30) **Francis**, J.A., and E. Hunter, 2007: Drivers of declining sea ice in the Arctic winter: A tale of two seas. *Geophysical Research Letters*, 34, L17503, doi:10:1029/2007GL030995.

31) **Meier**, M.F., M.B. Dyurgerov, U.K. Rick, S. O'Neel, W.T. Pfeffer, R.S. Anderson, S. Anderson, and A.F. Glazovsk, 2007: Glaciers dominate eustatic sea-level rise in the 21st century. *Science*, **317**(5841), 1064-1067.

32) **Rahmstorf**, S., 2007: A semi-empirical approach to projecting future sea-level rise. *Science*, **315**(5810), 368-370.

33) **Walter**, K.M., S.A. Zimov, J.P. Chanton, D. Verbyla, and F.S. Chapin, 2006: Methane bubbling from Siberian thaw lakes as a positive feedback to climate warming. *Nature*, **443(7107)**, 71-75.

THE U.S. CLIMATE CHANGE SCIENCE PROGRAM FOR FY 2009
CHAPTER REFERENCES (CONTINUED)

34) **Johansson**, T., and D. Lesnic, 2006: Reconstruction of a stationary flow from incomplete boundary data using iterative methods. *European Journal of Applied Mathematics*, **17(6)**, 651-663.

35) **Seager**, R., 2007: The turn of the century North American drought: Global context, dynamics, and past analogs. *Journal of Climate*, **20**(22), 5527-5552.

36) **Bates**, N.R., and A.J. Peters, 2007: The contribution of acid deposition to ocean acidification in the subtropical North Atlantic Ocean. *Marine Chemistry*, **107**, 547-558.

37) **Doney**, S.C., N. Mahowald, I. Lima, R.A. Feely, F.T. Mackenzie, J.-F. Lamarque, and P.J. Rasch, 2007: The impact of anthropogenic atmospheric nitrogen and sulfur deposition on ocean acidification and the inorganic carbon system. *Proceedings of the National Academy of Sciences*, **104**, 14580-14585, doi:10.1073/pnas.0702218104.

38) **CCSP**, 2007: *Effects of Climate Change on Energy Production and Use in the United States.* A Report by the U.S. Climate Change Science Program and the subcommittee on Global Change Research [Wilbanks, T. J., V. Bhatt, D.E. Bilello, S.R. Bull, J. Ekmann, W.C. Horak, Y.J. Huang, M.D. Levine, M.J. Sale, D.K. Schmalzer, and M.J. Scott (authors)]. Department of Energy, Office of Biological and Environmental Research, Washington, DC, USA, 160 pp.

39) **Bradley**, R.S., M. Vuille, H.F. Diaz, and W. Vergara, 2006: Threats to water supplies in the tropical Andes. *Science*, **312(5781)**, 1755-1756.

40) **CCSP**, 2007: *Scenarios of Greenhouse Gas Emissions and Atmospheric Concentrations* (Part A) and *Global-Change Scenarios: Their Development and Use* (Part B). A Report by the U.S. Climate Change Science Program and the Subcommittee on Global Change Research [Clarke, L., J. Edmonds, J. Jacoby, H. Pitcher, J. Reilly, R. Richels, E. Parson, V. Burkett, K. Fisher-Vanden, D. Keith, L. Mearns, C. Rosenzweig, M. Webster (authors)]. Department of Energy, Office of Biological and Environmental Research, Washington, DC, USA, 260 pp.

41) **Snover**, A.K., L. Whitely Binder, J. Lopez, E. Willmott, J. Kay, D. Howell, and J. Simmonds, 2007: *Preparing for Climate Change: A Guidebook for Local, Regional, and State Governments.* In association with and published by ICLEI – Local Governments for Sustainability, Oakland, CA. Available at <www.cses.washington.edu/cig/fpt/guidebook.shtml>.

HIGHLIGHTS OF
RECENT RESEARCH AND
PLANS FOR FY 2009

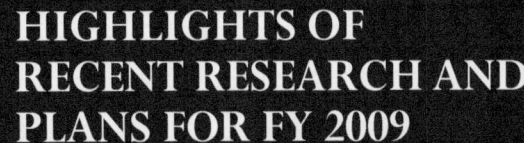

1 | Atmospheric Composition

Strategic Research Questions

3.1 What are the climate-relevant chemical, microphysical, and optical properties, and spatial and temporal distributions, of human-caused and naturally occurring aerosols?

3.2 What are the atmospheric sources and sinks of the greenhouse gases other than CO_2 and the implications for the Earth's energy balance?

3.3 What are the effects of regional pollution on the global atmosphere and the effects of global climate and chemical change on regional air quality and atmospheric chemical inputs to ecosystems?

3.4 What are the characteristics of the recovery of the stratospheric ozone layer in response to declining abundances of ozone-depleting gases and increasing abundances of greenhouse gases?

3.5 What are the couplings and feedback mechanisms among climate change, air pollution, and ozone layer depletion, and their relationship to the health of humans and ecosystems?

See Chapter 3 of the *Strategic Plan for the U.S. Climate Change Science Program* for detailed discussion of these research questions.

The Earth's atmosphere is composed of gases and particles that vary with spatial scale and with time, influencing climate, air quality, the stratospheric ozone layer, and weather. Interactions between these components have impacts on human health and the vitality of ecosystems and hence have high relevance to society. CCSP research on atmospheric composition focuses primarily on how human activities and natural processes affect atmospheric composition, and how these changes in turn relate to societal issues. The issues embrace multiple disciplines, cross many spatial scales, and are highly interrelated. Consequently, CCSP research is a highly coordinated endeavor that involves observational studies, laboratory investigations, and modeling analyses to provide the timely, accurate, and useful scientific information needed by decisionmakers nationally and internationally.

Progress has been made to date in understanding the role of atmospheric composition in Earth's climate. Efforts have been dedicated to the areas of largest uncertainty in understanding how atmospheric constituents other than carbon dioxide (CO_2) affect the forcing of climate. Atmospheric fine particles ("aerosols") can have either warming or cooling effects, depending on many factors. CCSP research has made progress in defining those factors, and has recently taken steps to address the next levels of complexity in the issue by looking at the interactions of aerosols with clouds. For FY 2009, CCSP's atmospheric composition research will focus on aerosols and aerosol/cloud interactions in the polar environment through analyses of measurements from satellites, aircraft, and the surface made during the International Polar Year (spring 2007 to summer 2008). Additional FY 2009 work focuses on linked air quality-climate modeling systems, future emission projections, and communicating research results to air quality decisionmakers.

HIGHLIGHTS OF RECENT RESEARCH

The following paragraphs provide selected highlights of recent research supported by CCSP-participating agencies.

Climate-Relevant Properties of Aerosols

Aerosol Forcing Effects on Climate Change are Better Defined.[1,2] Aerosols (atmospheric fine particles such as pollution, smoke, and desert dust) in sunlight directly heat the atmosphere if they absorb light, and cool the surface by absorbing and scattering light. Evaluating the net radiative effect of aerosols has been a key uncertainty in past Intergovernmental Panel on Climate Change (IPCC) assessments, partly due to the highly variable horizontal and vertical distribution of aerosol particles of differing chemical composition, size, and shape. This in turn hampers modeling efforts to understand the total amount and vertical distribution of solar radiation indicated by satellite observations. A recent field campaign studied the radiative forcing of atmospheric brown clouds—large pollution plumes that increasingly cover large regions of dry season southern Asia (see Figure 3). Critical measurements of solar heating profiles above, within, and below these pollution-dominated plumes over the Indian Ocean in the Northern Maldives were made with small unmanned aerial vehicles (UAVs) instrumented with aerosol and radiation detectors and positioned to make vertically aligned measurements. Findings indicated that continental air masses with higher aerosol particle concentrations, and in particular those containing the carbonaceous component of soot, exhibited increased aerosol absorption and heating by as much as

Figure 3: Ganges Valley Brown Cloud. Ganges Valley brown cloud plume drifts out over the Bay of Bengal and the Indian Ocean. *Credit: NASA / Goddard Space Flight Center.*

50% over background oceanic conditions. Other improvements in understanding have been made using the Aerosol Robotic NETwork (AERONET), a network of about 230 automated surface instruments that measure the optical properties of aerosols. Analysis of AERONET data collected in 2004 in the United Arab Emirates is leading to a better understanding of the dynamics of desert dust and pollution aerosols over a variety of environments in the Arabian Peninsula and over the Persian Gulf.

Current Trends in Arctic Haze and Implications for Climate Forcing. [1] The long-range transport of anthropogenic pollution from North America, Europe, and western Asia creates the aerosols associated with so-called Arctic haze. U.S., Finnish, and Canadian researchers have recently compiled long-term data to determine trends in and climate impacts of Arctic haze. The analysis confirmed previously reported results of a decreasing trend in Arctic haze between the early 1980s and the mid-1990s. In addition, the analysis revealed evidence of increasing levels of aerosol scattering, black carbon, and nitrate over the past decade. Calculations of direct radiative forcing by Arctic haze for a representative case during the haze maximum (mid-April) resulted in an estimated

2 to 3 Wm^{-2} of additional heating to the atmosphere and approximately 1 Wm^{-2} of cooling at the surface. CCSP researchers, as part of the International Polar Year, returned to the Arctic in 2008 using aircraft and ship platforms to better characterize the direct and indirect climate impacts of the Arctic haze.

Cloud-Aerosol-Climate Feedbacks and Interactions

Improved Understanding of Connections Between Aerosol Chemistry, Clouds, and Climate.[4,5,6,7] The increasing levels of aerosol from human activities affect cloud properties, cloud lifetime, and precipitation processes, and hence climate. However, the relationships are not well understood and the aerosol/cloud processes are one of the largest uncertainties in current understanding of climate change. Aerosols affect cloud properties by serving as cloud condensation nuclei (CCN) and thereby activating the formation of cloud droplets. The process of cloud drop activation is a function of both the size and composition of the aerosol particles, which, in turn, depend on the source of the aerosol and transformations that occur downwind. CCSP scientists conducted airborne and shipboard measurements of the aerosol number size distribution, aerosol chemical composition, and CCN concentration during the 2006 Gulf of Mexico Atmospheric Composition and Climate Study (GoMACCS). Aircraft measurements showed that the high organic emissions of the Houston Ship Channel region lead to a high organic acid content in the aerosol, and further revealed details of how the organic acid content evolves with increasing altitude within and above clouds to produce a ubiquitous layer of organic aerosol above cloud. Analysis of the GoMACCS shipboard data showed that when the organic content of the aerosol increases, the aerosol is less likely to form CCN and hence less likely to activate the formation of cloud droplets. These studies have led to an improved understanding of the links between aerosol composition and cloud drop formation. The work has yielded a simplified means of representing the processes in climate models, ultimately contributing to the development of an improved predictive capability. In the continental United States, 8 years of surface measurements at the Atmospheric Radiation Monitoring (ARM) Southern Great Plains

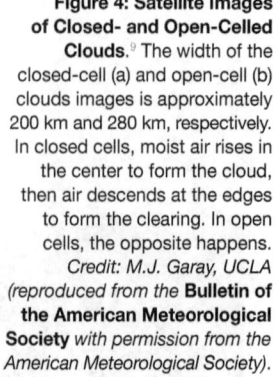

Satellite Images of Closed- and Open-Celled Clouds

site in Oklahoma have advanced understanding of how clouds, water vapor, and climate interact. The data have clearly shown that higher cloud fractions are the key factor in reducing the amount of solar radiation at the Earth's surface, whereas the amount of precipitable water vapor in the clouds has a much greater impact on the cloud's absorption of outgoing infrared radiation.

Atmospheric Aerosol Pollution Affects Structure and Reflectivity of Clouds.[8] Viewed from space, stratocumulus clouds can have either an "open cell" or a "closed cell" appearance to their fine-scale structure (see Figure 4). The structure greatly affects the degree to which the clouds reflect light and, hence, the climate-relevant properties of the clouds. Recently it was hypothesized that precipitation may trigger the transition from closed to open cellular structure. Precipitation tends to occur in clean regions lacking in aerosol, thereby providing a potential link between the composition of the atmosphere and the organization of clouds. Polluted, non-precipitating clouds were shown to exhibit a closed cellular structure, whereas in clean conditions, open cells formed in response to drizzle. CCSP researchers have modeled the processes and confirmed the hypothesis. The transition from closed to open cells has dramatic implications for radiative forcing, essentially representing the transition from a highly reflective cloud to one of much lower reflectance.

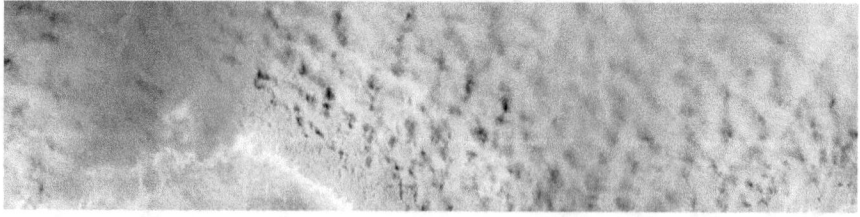

Atmospheric Constituents other than CO_2, including Water Vapor, and Implications for Earth's Energy Balance

Lightning and Pollution Combine to Cause Ozone Enhancements in the Summer Upper Troposphere.[10,11,12,13,14] In the upper troposphere, ozone acts as a greenhouse gas and hence is relevant to climate. Analyzing data from a summer 2004 study, CCSP researchers found unexpectedly high levels of ozone in the upper troposphere 10 to 11 km above eastern North America during summer that were not attributable to either the high amounts of ozone pollution at Earth's surface or the higher ozone levels in the stratosphere. The researchers investigated the cause of the increased ozone, which can nearly double the amount of upper-tropospheric ozone above the region. It was found that a natural factor—the emission of nitrogen oxides from lightning—acts in concert with the generally higher background levels of ozone precursor compounds in today's polluted atmosphere to produce much of the upper-tropospheric ozone enhancement (see Figure 5). This ozone enhancement with a strong natural component contributes

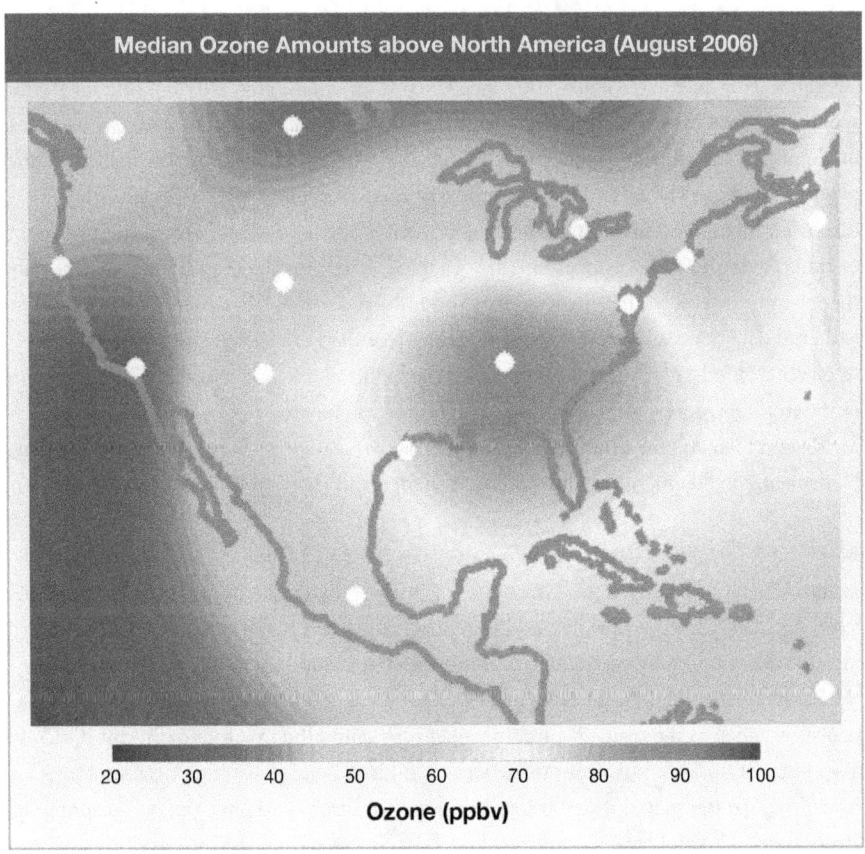

Median Ozone Amounts above North America (August 2006)

Ozone (ppbv)

Figure 5: **Median Ozone Amounts above North America (August 2006)**. Median ozone amounts above North America, in parts per billion, at 10 to 11 km during August 2006, after stratospheric ozone contribution was removed. The ozone enhancement is mainly due to nitrogen oxides emitted by lightning into the upper troposphere, followed by reactions with carbon monoxide, methane, and volatile organic compounds of both anthropogenic and natural origins. The white dots show the locations of observing stations. *Credit: O.R. Cooper, CIRES and NOAA / Earth System Research Laboratory (reproduced from* **Journal of Geophysical Research** *with permission from the American Geophysical Union).*

to the radiation budget at the regional scale. Other data gathered in the study demonstrate that the influence of lightning and convection on upper-tropospheric ozone has been previously underestimated. Lightning may have been underestimated by a factor of four, and faster convection rates may be needed to accurately model this region of the atmosphere. Large differences remain between observed and modeled levels of free radicals in the upper troposphere, an indicator that a major uncertainty remains in the understanding of how lightning-produced nitrogen oxides affect ozone in the upper troposphere.

Field Mission Investigates Atmospheric Composition, Clouds, and Climate in the Tropical Atmosphere. CCSP researchers carried out a field mission that gathered chemical and meteorological data for the cold, dry conditions of the upper tropical tropopause (an important transition region between the troposphere and stratosphere). This region of the Earth's atmosphere between 14 and 18 km plays a key role in both climate change science and depletion of the stratospheric ozone layer. The Tropical Composition, Clouds, and Climate Coupling (TC4) experiment, based in San Jose, Costa Rica, involved dozens of scientists from U.S. agencies and academia and used a unique combination of coordinated observations from satellites, ground stations in the inter-tropical convergence zone, and instrumented aircraft during July and August 2007 (see <www.espo.nasa.gov/tc4/>). One of the specific goals of TC4 was to study the composition, formation, and radiative properties of clouds (cirrus and sub-visible cirrus) in this region, thereby assessing the contributions of such clouds, aerosols, and water vapor to climate forcing. Other aspects of the campaign focused on understanding the convective processes that control the transport of trace gases and aerosols from the lower atmosphere into the tropical tropopause and thence into the stratosphere, where they can influence stratospheric ozone. This mission gathered data that CCSP researchers are using to expand the scientific understanding of climate-cloud-chemistry interactions in the highly active and climate-relevant region of the tropical tropopause. The measurements also will be used to test retrieval algorithms for several instruments on the Aura satellite that measure trace gases (High Resolution Dynamic Limb Sounder, Microwave Limb Sounder, and Thermal Emission Spectrometer).

Global Transport of Pollution from Satellite Observations of Carbon Monoxide.[15,16] **Carbon** monoxide (CO) in the Earth's atmosphere is formed by the incomplete combustion of fossil fuels and biomass burning, and is primarily emitted at the surface. It can be lifted into the atmosphere by convection and transported around the globe by the prevailing winds. Its relatively long lifetime enables CO to be a good tracer of transport processes, such as the trans-Pacific transport of Asian pollutants to North America. Two years of observations from the Microwave Limb Sounder on the Aura satellite have provided the spatial distribution, temporal variation, and long-range transport of atmospheric CO and have shown the close relationship of concentrations of this gas to

surface emissions, deep convection, and horizontal winds. The transport of CO from Southeast Asia across the Pacific to North America occurs most frequently during the Northern Hemisphere summer when deep convection associated with the Asian monsoon is clustered over the strong anthropogenic emission regions. Measurements of the global distribution of CO over time provide a strong indicator of the connections between changes in air quality due to increased industrialization and climate.

Regional Pollution and Global Climate Change

Impact of Global Change on U.S. Regional Air Quality.[17,18,19,20] Emerging study results show that climate change has the potential to increase ground-level ozone concentrations in many areas of the United States and to lengthen the season of elevated ozone events. These increases may be beyond the envelope of natural interannual variability. Planned and future emissions controls will lower U.S. ozone concentrations, but the impacts of global change may necessitate further reductions to meet national air quality standards.

Recovery of the Stratospheric Ozone Layer

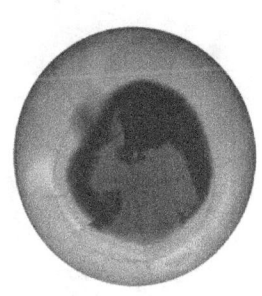

A New Formulation for Gauging the Effects of Ozone-Depleting Substances on the Ozone Layer.[21] Equivalent effective stratospheric chlorine (EESC) is a convenient parameter to quantify the effects of halogens (chlorine and bromine) on the depletion of the stratospheric ozone layer. EESC has been extensively used to evaluate future scenarios of ozone-depleting substances (ODS) in the stratosphere. CCSP research has led to a new formulation of EESC that provides revised estimates of ozone layer recovery. The work shows that ODS levels will recover to 1980 levels in the year 2041 in the mid-latitudes, and 2067 over Antarctica, assuming adherence to international agreements that regulate the use of ODS. The researchers also assessed the uncertainties in the estimated recovery times. The 95% confidence interval associated with the mid-latitude recovery is the time period from 2028 to 2049, while it is from 2056 to 2078 for Antarctic ODS recovery.

HIGHLIGHTS OF PLANS FOR FY 2009

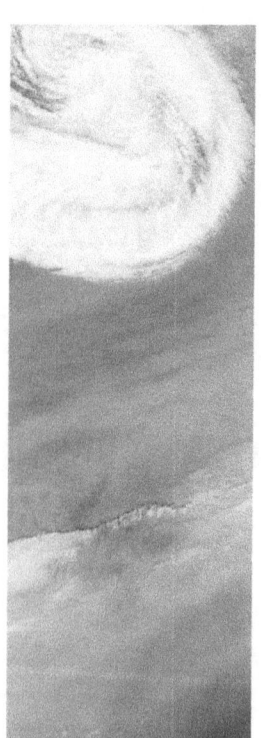

CCSP will continue to gather and analyze information through measurement, modeling, and assessment studies to enhance understanding of atmospheric composition and of the processes affecting atmospheric chemistry. The general emphasis will continue to be on quantifying the effects of aerosols and non-CO_2 greenhouse gases on climate. In FY 2009, the following research activities will be emphasized to meet the overall priority.

Analysis of International Polar Year Data Sets. **Research** across several agencies will focus on analyzing data sets gathered during the 2008 spring and summer experiments to study Arctic aerosols and their connections to clouds, radiation, and ice melting. Measurements from satellites, aircraft, and the surface were made to assess the influences of long-range pollution transported to the Arctic environment, the so-called Arctic haze. Specific areas of scientific investigation include the effects of Arctic haze on ozone chemistry and ice melting; studies of the influence of boreal forest fires on the Arctic climate; and an evaluation of aerosol/cloud influences in the Arctic.

This activity will address Questions 3.1 and 3.3 of the CCSP Strategic Plan.

Utilization of Ground-Based Measurement Networks for Climate and Ozone Studies. **Retrievals** of atmospheric particulate absorption from AERONET will continue to be utilized in climate forcing studies and in the validation of current and future satellite missions, such as the Glory mission (2009 launch) that will measure sunlight absorption by aerosols. Network expansion will continue, with a focus on inadequately sampled regions that are important for understanding global climate change, such as China (both the polluted eastern regions and the western deserts that are a source of dust storms). An experimental effort is underway to investigate the sensitivity of ground-based instruments for nighttime measurements using moonlight. Studies using the Micro Pulse Lidar Network data will focus on the influence of polar stratospheric clouds on ozone destruction over the Antarctic and on the impacts of Arctic haze on polar climate.

This activity will address Questions 3.1 and 3.3 of the CCSP Strategic Plan.

Continue Regional Aerosol Study in China. **In 2008,** the ARM Mobile Facility (AMF) was deployed in China to identify and quantify the climatic effects of aerosols. The AMF was deployed at Lake Taihu (April-December) near Shanghai, China, and the ARM Ancillary Facility was also deployed from Linze (February-June) and Xianghe (July-December). Data will be used to improve rain remote sensing and understanding the roles of aerosols in affecting regional climate and atmospheric circulation.

This activity will address Questions 3.1, 3.2, and 3.3 of the CCSP Strategic Plan.

Cloud/Aerosols Field Study. **Extensive** and persistent layers of stratus clouds occur off the subtropical west coasts of Africa and of North and South America. These cloud decks have a significant impact on Earth's radiation budget. Aerosols, arising from natural processes and from human activity, have important influences on the brightness and persistence of these clouds. The Variability of the American Monsoon System (VAMOS) Ocean-Cloud-Atmosphere-Land Study (VOCALS) field mission will study the stratus deck off the Pacific coast of Chile and Peru, using *in situ* and remote aircraft observations, along with satellite and ship-based measurements. Natural and human sources of particles will be observed, as well as the roles these particles play in the determining the brightness and lifetimes of stratus clouds. The VOCALS planning and site research occurred early in 2008, and the field mission will occur in October/November 2009.

This activity will address Questions 3.1, 3.2, and 3.3 of the CCSP Strategic Plan.

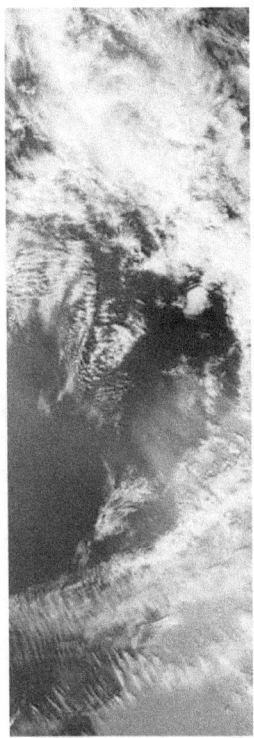

Examination and Intercomparison of Water Vapor Measurements from Aircraft, Balloons, and Satellites. Understanding changes in the distribution of water vapor, whether due to natural or anthropogenic causes, is essential to understanding the potential for climate change. Shortfalls in knowledge of the processes affecting water vapor concentrations near the interface between the troposphere and the stratosphere result primarily from the difficulties in making accurate water vapor measurements at these altitudes where concentrations are quite small. A number of research efforts will be continued or initiated to help resolve observed discrepancies in *in situ* water vapor observations. These activities are being conducted jointly by two CCSP agencies with the involvement of U.S. and international investigators from a wide range of government and academic institutions. The planned efforts include (1) single instrument laboratory studies designed to better characterize and understand instrument performance and calibration under a variety of atmospheric conditions; (2) the possible selection and use of a water vapor calibration standard to establish and/or confirm measurement accuracy and precision; and (3) multiple-instrument intercomparisons in the laboratory and field involving an independent referee to coordinate and present the results of each formal laboratory and flight intercomparison that includes instruments from different research groups.

This activity will address Questions 3.2 and 3.5 of the CCSP Strategic Plan.

Interactions of Climate Change and Air Quality. **Understanding** the combined effect of climate change and air quality is a key research question. Continuing work in FY 2009 will focus on (1) reducing the uncertainty for ground-level ozone; (2) assessing the impact of climate change on particulate matter; and (3) preliminary research to enable assessment of interactions with mercury. The FY 2009 work focuses on linked air quality-climate modeling systems, future emission projections, and communicating research results to air quality decisionmakers.

This activity will address Questions 3.3 and 3.5 of the CCSP Strategic Plan.

Aviation Impacts on the Upper Atmosphere and Climate Change. CCSP agencies concerned with understanding the impacts of aviation on climate change will address key uncertainties and research gaps through the Aviation-Climate Change Research Initiative (ACCRI). ACCRI will use a structured, sequential four-step approach, simultaneously applying the latest scientific knowledge and modeling/analysis capability to quantify the climate impacts of aviation. The main objective of ACCRI is to inform policymaking decisions for the U.S. Next Generation Air Transportation System and the International Civil Aviation Organization Committee on Aviation Environmental Protection based on the latest research results. A key goal of ACCRI is to identify, develop, and evaluate the metric(s) that can capture aviation-induced climate impacts at all relevant spatial and temporal scales and assist in tradeoff analysis whenever possible.

This activity will address Questions 3.3 and 3.5 of the CCSP Strategic Plan.

ATMOSPHERIC COMPOSITION
CHAPTER REFERENCES

1) **Eck**, T.F., B.N. Holben, J.S. Reid, A. Sinyuk, O. Dubovik, A. Smirnov, D. Giles, N.T. O'Neill, S.-C. Tsay, Q. Ji, A. Al Mandoos, M. Ramzan Khan, E.A. Reid, J.S. Schafer, M. Sorokine, W. Newcomb, and I. Slutsker, 2008: Spatial and temporal variability of column-integrated aerosol optical properties in the southern Arabian Gulf and United Arab Emirates in summer. *Journal of Geophysical Research*, **113**, D01204, doi:10.1029/2007JD008944.

2) **Ramanathan**, V., M.V. Ramana, G. Roberts, D. Kim, C. Corrigan, C. Chung, and D. Winker, 2007: Warming trends in Asia amplified by brown cloud solar absorption. *Nature*, **448**, 575-578, doi:10.1038/nature/06019.

3) **Quinn**, P.K., G. Shaw, E. Andrews, E.G. Dutton, T. Ruoho-Airola, and S.L. Gong, 2007: Arctic haze: current trends and knowledge gaps. *Tellus*, **59B**, 99-114.

4) **Dong**, X., B. Xi, and P. Minnis, 2006: Observational evidence of changes in water vapor, clouds, and radiation at the ARM SGP site. *Geophysical Research Letters*, **33**, L19818, doi:10.1029/2006GL027132.

5) **Quinn**, P.K., T.S. Bates, D.J. Coffman, and D.S. Covert, 2007: Influence of particle size and chemistry on the cloud nucleating properties of aerosols. *Atmospheric Chemistry and Physics Discussion*, **7**, 14171-14208.

6) **Sorooshian**, A., M.-L. Lu, F.J. Brechtel, H. Fonsson, G. Feingold, R.C. Flagan, and J.H. Seinfeld, 2007: On the source of organic acid aerosol layers above clouds. *Environmental Science and Technology*, **41**, 4647-4654.

7) **Sorooshian**, A., N.L. Ng, A.W.H. Chan, G. Feingold, R.C. Flagan, and J.H. Seinfeld, 2007: Particulate organic acids and overall water-soluble aerosol composition measurements from the 2006 Gulf of Mexico Atmospheric Composition and Climate Study. *Journal of Geophysical Research*, **112**, D13201, doi:10.1029/2007JD008537.

8) **Xue**, H., G. Feingold, and B. Stevens, 2008: Aerosol effects on clouds, precipitation, and the organization of shallow cumulus convection. *Journal of Atmospheric Science*, **65**, 392-406, doi:10.1175/2007JAS2428.1.

9) **Garay**, M.J., R. Davies, C. Averill, and J.A. Westphal, 2004: Actinoform clouds: Overlooked examples of cloud self-organization at the mesoscale. *Bulletin of the American Meteorological Society*, doi:10.1175/BAMS-85-10-1585.

ATMOSPHERIC COMPOSITION
CHAPTER REFERENCES (CONTINUED)

10) **Cooper**, O.R., A. Stohl, M. Trainer, A.M. Thompson, J.C. Witte, S.J. Oltmans, G. Morris, K.E. Pickering, J.H. Crawford, G. Chen, R.C. Cohen, T.H. Bertram, P. Wooldridge, A. Perring, W.H. Brune, J. Merrill, J.L. Moody, D. Tarasick, P. Nédélec, G. Forbes, M.J. Newchurch, F.J. Schmidlin, B.J. Johnson, S. Turquety, S.L. Baughcum, X. Ren, F.C. Fehsenfeld, J.F. Meagher, N. Spichtinger, C.C. Brown, S.A. McKeen, I.S. McDermid, and T. Leblanc, 2006: Large upper tropospheric ozone enhancements above midlatitude North America during summer: In situ evidence from the IONS and MOZAIC ozone measurement network. *Journal of Geophysical Research*, **111**, D24S05, doi:10.1029/2006JD007306.

11) **Cooper**, O.R., M. Trainer, A.M. Thompson, S.J. Oltmans, D.W. Tarasick, J.C. Witte, A. Stohl, S. Eckhardt, J. Lelieveld, M.J. Newchurch, B.J. Johnson, R.W. Portmann, L. Kalnajs, M.K. Dubey, T. Leblanc, I.S. McDermid, G. Forbes, D. Wolfe, T. Carey-Smith, G.A. Morris, B. Lefer, B. Rappenglück, E. Joseph, F. Schmidlin, J. Meagher, F.C. Fehsenfeld, T.J. Keating, R.A. Van Curen, and K. Minschwaner, 2007: Evidence for a recurring eastern North America upper tropospheric ozone maximum during summer. *Journal of Geophysical Research*, **112**, D23304, doi:10.1029/2007JD008710.

12) **Hudman**, R.C., D.J. Jacob, S. Turquety, E.M. Leibensperger, L.T. Murray, S. Wu, A.B. Gilliland, M. Avery, T.H. Bertram, W. Brune, R.C. Cohen, J.E. Dibb, F.M. Flocke, A. Fried, J. Holloway, J.A. Neuman, R. Orville, A. Perring, X. Ren, G.W. Sachse, H.B. Singh, A. Swanson, and P.J. Wooldridge, 2007: Surface and lightning sources of nitrogen oxides in the United States: Magnitudes, chemical evolution, and outflow. *Journal of Geophysical Research*, **112**, D12S05, doi:10.1029/2006JD007912.

13) **Bertram**, T.H., A.E. Perring, P.J. Wooldridge, J.D. Crounse, A.J. Kwan, P.O. Wennberg, E. Schauer, J. Dibb, M. Avery, G. Sachse, S.A. Vay, J.H. Crawford, C.S. McNaughton, A. Clark, K.E. Pickering, H. Fuelberg, G. Huey, D.R. Blake, H.B. Singh, S.R. Hall, R.E. Shetter, A. Fried, B.G. Heikes, and R.C. Cohen, 2008. Direct Measurements of the convective recycling of the upper troposphere. *Science*, **315(5813)**, 816-820, doi:10.1126/science.1134548.

14) **Ren**, X., J.R. Olson, J.H. Crawford, W.H. Brune, J. Mao, R.B. Long, Z. Chen, G. Chen, M.A. Avery, G.W. Sachse, J.D. Barrick, G.S. Diskin, L.G Huey, A. Fried, R.C. Cohen, B. Heikes, P. Wennberg, H.B. Singh, D.R. Blake, and R.E. Shetter, 2008: HO$_x$ observation and model comparison during INTEX-A 2004: observation, model calculation, and comparison with previous studies. *Journal of Geophysical Research*, **113**, D05310, doi:10.1029/2007/JD009166.

15) **Jiang**, J.H., N.J. Livesey, H. Su, L. Neary, J.C. McConnell, and N.A.D. Richards, 2007: Connecting surface emissions, convective uplifting, and long-range transport of carbon monoxide in the upper troposphere: New observations from the Aura Microwave Limb Sounder. *Geophysical Research Letters*, **34**, L18812, doi:10.1029/2007GL030638.

16) **Schoeberl**, M.R., B.N. Duncan, A.R. Douglass, J. Waters, N. Livesey, W. Read, and M. Filipiak, 2006: The carbon monoxide tape recorder. *Geophysical Research Letters*, **33**, L12811, doi:10.1029/2006GL026178.

17) **Bell**, M.L., R. Goldberg, C. Hogrefe, P.L. Kinney, K. Knowlton, B. Lynn, J. Rosenthal, C. Rosenzweig, and J.A. Patz, 2007: Climate change, ambient ozone, and health in 50 U.S. cities. *Climatic Change*, **82**, 61-72, doi:10.1007/s10584-006-9166-7.

18) **Tagaris**, E., K. Manomaiphiboon, K.-J. Liao, L.R. Leung, J.-H. Woo, S. He, P. Amar, and A.G. Russell, 2007: Impacts of global climate change and emissions on regional ozone and fine particulate matter concentrations over the United States. *Journal of Geophysical Research*, **112**, D14312, doi:10.1029/2006JD008262.

19) **Tao**, Z., A. Williams, H.-C. Huang, M. Caughey, and X.-Z. Liang, 2007: Sensitivity of U.S. surface ozone to future emissions and climate changes. *Geophysical Research Letters*, **34**, L08811, doi:10.1029/2007GL029455.

20) **Wu**, S., L.J. Mickley, E.M. Leibensperger, D.J. Jacob, D. Rind, and D.G. Streets, 2008: Effects of 2000-2050 global change on ozone air quality in the United States. *Journal of Geophysical Research*, **108**, D06302, doi:10.1029/2007JD008917.

21) **Newman**, P.A., J.S. Daniel, D.W. Waugh, and E.R. Nash, 2007. A new formulation of equivalent effective stratospheric chlorine (EESC). *Atmospheric Chemistry and Physics*, **7**, 4537-4552.

2 | Climate Variability and Change

Strategic Research Questions

4.1 To what extent can uncertainties in model projections due to climate system feedbacks be reduced?

4.2 How can predictions of climate variability and projections of climate change be improved, and what are the limits of their predictability?

4.3 What is the likelihood of abrupt changes in the climate system such as the collapse of the ocean thermohaline circulation, inception of a decades-long mega-drought, or rapid melting of the major ice sheets?

4.4 How are extreme events, such as droughts, floods, wildfires, heat waves, and hurricanes, related to climate variability and change?

4.5 How can information on climate variability and change be most efficiently developed, integrated with non-climatic knowledge, and communicated in order to best serve societal needs?

See Chapter 4 of the *Strategic Plan for the U.S. Climate Change Science Program* for detailed discussion of these research questions.

To address fundamental CCSP goals, the climate variability and change (CVC) element emphasizes research to improve descriptions and understanding of past and current climate, as well as to advance national modeling capabilities to simulate climate and project how climate and related Earth systems may change in the future. Research under this element encompasses time scales ranging from short-term climate variations of a season or less to longer term climate changes occurring over decades to centuries. The CVC element places a high priority on improving understanding and predictions of phenomena that may cause high impacts on society, the economy, and the environment. Examples include identifying the relationships between variations and changes in climate and hurricane activity; improving understanding and predictions of droughts; increasing understanding of and capabilities to predict the El Niño-Southern Oscillation and its attendant impacts; identifying processes that may produce rapid or accelerated climate change; and improving capabilities to observe, understand, and model Earth system

components that have high societal and environmental relevance, including sea ice, glaciers, ice sheets, and sea level. Addressing these fundamental issues requires an integrated approach toward understanding the interactions and feedbacks among the different components of the Earth system, including the atmosphere, ocean, land, cryosphere, and biosphere.

CVC research is placing increasing emphasis on understanding and modeling the links and feedbacks among climate system components. Considerable advances have been made in this area through the development and application of Earth system models, including several that were used extensively in the recently completed Intergovernmental Panel on Climate Change (IPCC) Fourth Assessment Report (see introductory chapter of this report). CVC research is also emphasizing the development of new capabilities to link Earth system models together with Earth system observations to produce internally consistent maps of atmospheric, oceanic, land surface, and ice conditions that are called "Earth system analyses." These analyses will help us to understand and explain past and current climate conditions, and provide decisionmakers with new tools to track how the Earth system is evolving in time over the entire planet.

Research within the CVC element focuses on two broad, critically important questions to society defined in the CCSP Strategic Plan:
- How are climate variables that are important to human and natural systems affected by changes in the Earth system resulting from natural processes and human activities?
- How can emerging scientific findings on climate variability and change be further developed and communicated in order to better serve societal needs?

More specifically, CVC research addresses the five strategic research questions listed at the beginning of this chapter to achieve the milestones, products, and payoffs described in the CCSP Strategic Plan. Cooperative efforts involving CCSP agencies have led to significant progress in addressing the strategic questions articulated in the CVC chapter of the CCSP Strategic Plan. The following section highlights some of the major scientific advances achieved during this past fiscal year.

HIGHLIGHTS OF RECENT RESEARCH

This year's highlights of climate variability and change research emphasize advances in several areas of high societal interest. These include improving understanding of tropical cyclone-climate links; droughts; the predictability of the El Niño-Southern Oscillation; changes in the North Atlantic Ocean circulation; ice processes and implications for sea-level changes; the causes of sea ice declines; and detection and attribution of the causes for regional climate changes, including effects of air pollution on Asian climate.

Assessing Tropical Cyclone-Climate Links from Paleoclimate Data.[1,2] A new line of research is applying paleoclimate data to study tropical cyclone behavior before the period of modern observational records, a field called "paleotempestology." This research provides the potential to study climate-tropical cyclone relationships through a wider range of climates than can be obtained from the modern observational period. By greatly increasing the time interval under consideration, this approach also enables the possibility of detecting rare, catastrophic events that have very long recurrence intervals (hundreds of years or longer). Using marine sediment cores and corals, two recent studies have quantified the frequency of major hurricanes in the North Atlantic over the past 5,000 years. They find that the frequency of major hurricanes has varied on centennial to millennial time scales; however, a simple relationship to sea surface temperature variations seems to be lacking. These studies suggest the potential importance of other factors, such as variations in the El Niño-Southern Oscillation, tropical monsoons, and trade wind strength in determining variations in the strength of hurricanes.

Climate-Hurricane Links Over the North Atlantic.[3,4,5,6] Vigorous scientific debate continues about the detection of trends in Atlantic hurricane activity, with several studies published within the past year giving conflicting findings. One recent study showed a strong positive correlation between multi-decadal sea surface temperature variations and tropical cyclone counts and concluded that a trend toward increasing numbers of tropical cyclones in the Atlantic basin has been driven predominantly by anthropogenically forced increases in Atlantic sea surface temperatures. In contrast, another study suggested that the estimated increase in the number of Atlantic hurricanes over the 20th century is mostly, if not entirely, due to observational uncertainties that lead to substantial storm undercounts early in the period. Yet another study argues that a trend toward increasing Atlantic hurricane activity is insensitive to a range of estimates of storm undercounts. Efforts continue to reanalyze

the tropical cyclone record, as well as the
climate of the early part of the 20th
century. This work should improve ability
to describe and understand the causes of
climate-hurricane relationships.

*Modeling Studies of Climate Factors Influencing
Hurricanes.*[7,8] Recent research on Atlantic
hurricanes and climate change has focused on whether the increase in hurricane activity
in the basin since the 1970s portends future large increases in a warming climate. In an
analysis of projected climate changes over the tropical Atlantic region during the 21st
century derived from 18 different climate models developed for the IPCC Fourth
Assessment Report, a notable finding was that the vertical wind shear (the difference
in wind direction and speed between the lower and upper atmosphere) is projected to
increase across much of the Caribbean in the warmer climate, a factor that tends to
suppress tropical storm and hurricane development and intensification. In other basins,
shear was reduced, which would tend to favor enhanced tropical storm activity. This
study emphasizes the importance of incorporating changes in vertical wind shear as
well as sea surface temperatures in projections of future hurricane activity.

A second study introduced a new regional modeling approach for understanding
mechanisms that produce changes in Atlantic hurricane activity. The study uses a high-
resolution model (18-km, or about 11-mile grid spacing) in which the large-scale
atmospheric conditions are provided from historical analyses ("reanalyses") and ocean
conditions are derived from observed sea surface temperatures. Within the evolving
large-scale state, the model generates hurricane activity. This method successfully models
the observed increase in Atlantic hurricane activity over the period 1980 to 2006, as
well as other features like diminished hurricane activity during El Niño years. This new
approach should provide an important means for addressing key questions regarding
possible responses of Atlantic hurricane activity to future climate warming.

Climate Model Projections of Increasing Aridity in Southwestern North America.[9] A recent study
examines how the hydroclimate over southwestern North America may change over
the remainder of this century. The study evaluates climate change projections out to
2100 derived from 19 climate models used in the IPCC Fourth Assessment Report.
These models were forced with the A1B emissions scenario, in which carbon dioxide
(CO_2) emissions increase until about 2050 and then decrease modestly. The study
shows that 18 of the 19 models predict a drier climate for the southwest United States.
The projected trend in "precipitation minus evaporation," an indicator of drought
potential (see Figure 6), implies that droughts as severe as the 1930s Dust Bowl
drought may become common occurrences in this region within the next few decades.

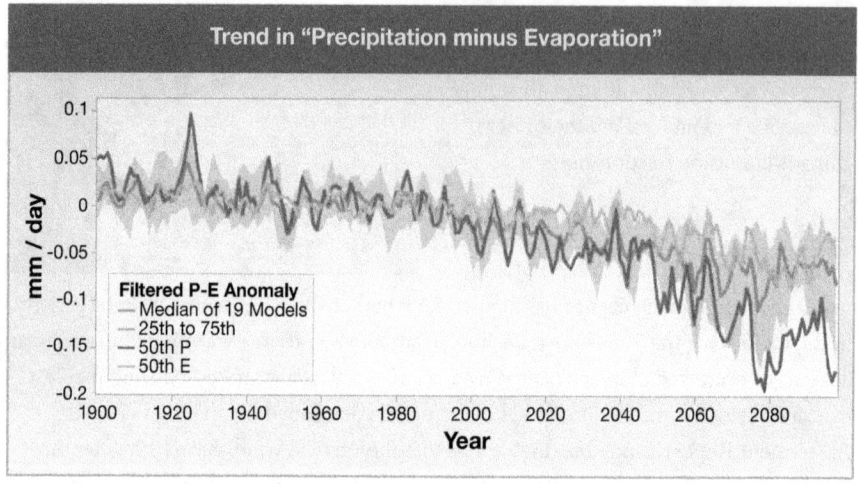

Figure 6: Trend in "Precipitation minus Evaporation". The trend in "precipitation minus evaporation" for the southwest United States simulated by the climate models used for the IPCC Fourth Assessment Report. *Credit: R. Seager, Lamont Doherty Earth Observatory (reproduced from* **Science** *with permission from the American Association for the Advancement of Science).*

The projections suggest that rain-producing storms that normally affect the southwest United States will shift northward, making the region drier. Further work will be required to understand the reason for the shifts in the storm patterns.

Seasonal-to-Interannual Predictability of the Tropical Pacific and Atlantic.[10,11,12] A debate currently exists about the extent to which the predictability of the El Niño-Southern Oscillation (ENSO) is limited by uncertainties in specifying initial conditions or by errors associated with "weather noise" and other unresolved physical processes. Understanding what limits climate predictability is critical to quantifying confidence in ENSO predictions and related impacts over North America. Recent research, using a new interactive ensemble coupling strategy, suggests the skill of current operational ENSO forecasts is limited mainly by initial condition errors. Other sources of error, however, do have a significant impact on the predictability of La Niña events and large El Niño events. For the tropical Atlantic sector, research is exploring the impact of biases in

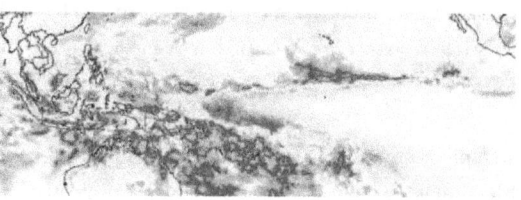

model simulations on their ability to reproduce and predict the leading patterns of climate variability in this region, with the goal of improving operational seasonal forecasts.

New Observations of the North Atlantic Ocean Circulation.[13,14] An issue of major concern is the potential for rapid and dramatic collapse of the northward flow of warm water in the North Atlantic, which may have serious implications for the climate of surrounding regions, including North America and Europe. A new international observing program,

partly supported by U.S. agencies, is examining whether and how this flow is changing. The northward flow of relatively warm, near-surface water is closely linked to a southward flow of cold, deep water, which forms an oceanic conveyer belt called the Atlantic Meridional Overturning Circulation (AMOC). Two recent studies report on observations of the variability in the AMOC, which is a fundamental prerequisite to understanding trends in this circulation, as well as its potential for rapid collapse. The studies indicate surprisingly large variability over the course of the 1-year period for this new program, suggesting that trends detected in previous studies from "snapshot" observations may be strongly influenced by the sampling of this large intra-annual variability. Observations from this program appear to be of sufficient accuracy to detect any large abrupt transition of the AMOC. The short period of record precludes a description of how the AMOC varies interannually. Understanding this variability will be critical for improving climate models used in predictions and projections of climate variability and change.

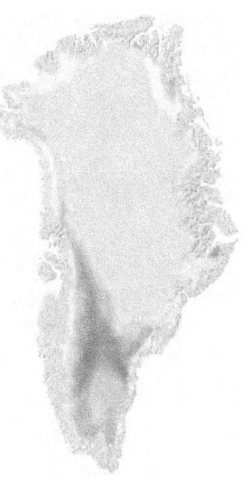

Ice Processes and Sea Level Change.[15,16,17] Future sea-level rise will depend on how much the ocean expands from future warming, as well as ice loss to the sea from glaciers, ice caps, the larger Greenland and Antarctic Ice Sheets, and other factors. A recent study found that about 60% of the ice loss in recent years has been from glaciers and ice caps (features whose perennial snow and ice coverage is less than 50,000 km²) rather than the larger ice sheets. Acceleration of glacier melt, in part due to processes not yet considered in models, may cause an additional 0.1 to 0.25 m of sea-level rise by 2100. An analysis of Special Sensor Microwave/Imager (SSM/I) radiometer measurements over the Greenland Ice Sheet back to 1988 shows that there was an extended period of melting snow in 2006, with snowmelt occurring more than 10 days longer than the 1988-2005 average over certain areas. Yet another separate analysis using satellite-derived surface elevation data found major variations in recent ice discharge and mass loss at two of Greenland's largest outlet glaciers. These studies point to the need to better constrain observations and advance understanding of ice outlet dynamics in order to improve projections of future sea-level rise. As noted in the IPCC Fourth Assessment Report, there are presently large uncertainties about the processes controlling ice flows, and advances in this area will be critical to improving models required for estimating future sea-level rise.

Declines in Arctic Sea Ice.[18,19] There have been dramatic declines in Arctic sea ice in recent decades that are particularly strong in summer and early fall (Figure 7), but are evident throughout the year. A study compared observations against sea ice simulations from more than a dozen models that were used in the IPCC Fourth Assessment Report. The results indicated that nearly all the models failed to produce as much of a decrease in minimum Arctic sea ice coverage as was observed, often substantially underestimating

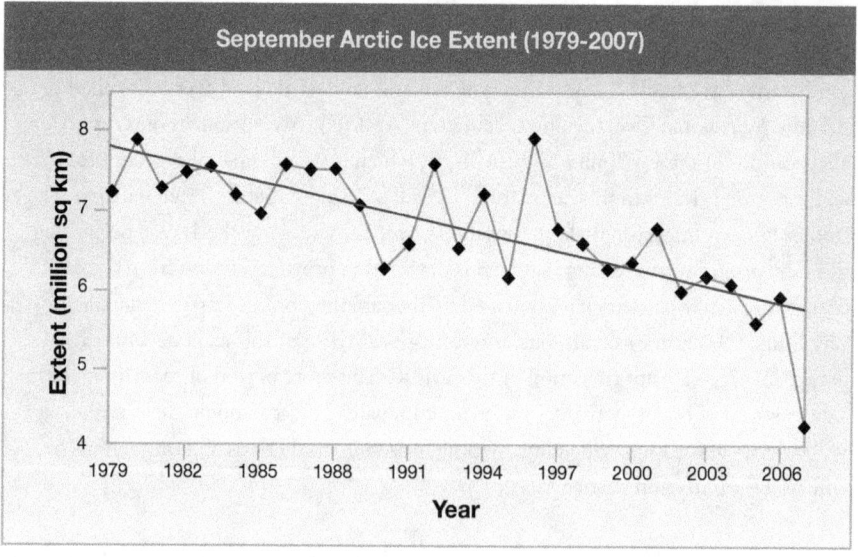

September Arctic Ice Extent (1979-2007)

the decreases. The authors suggest that the Arctic could be seasonally free of sea ice earlier than IPCC projections. Another study considered factors that may contribute to declines in sea ice in the Arctic winter. The results show important regional differences in the processes leading to the changes, although the hemispheric-mean decline in winter ice extent appears primarily due to increasing sea surface temperatures in the Barents Sea and adjoining waters. This result is in contrast to earlier work showing that downwelling longwave radiation is the leading factor contributing to changes in Arctic minimum sea ice extent in late summer and early fall, although relative contributions from different factors vary by region and year. The results emphasize the important roles of both atmospheric and oceanic processes in understanding and modeling the causes of rapid sea ice declines.

Detection of Human Influence on 20th Century Precipitation Trends.[20,21,22] Human influence on climate has been detected in numerous quantities, but changes in precipitation have been more difficult to attribute to human causes, partly due to sampling errors associated with inferring global precipitation trends from the archive of mostly land-based measurements. Two recent studies shed new light on this detection problem. One study makes use of monthly precipitation observations over global land areas for 50 to 75 years prior to 2000 and demonstrates that anthropogenic forcing has had a detectable influence on observed changes in average precipitation within latitudinal bands. In particular, anthropogenic forcing has contributed significantly to observed increases in precipitation in the Northern Hemisphere mid-latitudes, drying in the Northern Hemisphere subtropics and tropics, and moistening in the Southern Hemisphere subtropics and deep tropics. These opposing zonal trends help explain why detecting and explaining a net global trend has been so challenging. Satellite observations of precipitation reduce sampling errors but are available for a more limited period of time. In a second study, satellite observations are used to show a globally averaged increase in evaporation, precipitation, and total water of about 1.3% per decade for the period July 1987 to August 2006. During that time period, satellite measurements showed a positive trend in the globally averaged surface and lower atmosphere temperatures of about 0.2°C per decade, implying an increase in evaporation, precipitation, and total water of about 6.5% per °C. These increases in evaporation and total water agrees with climate models, but the models predict that global precipitation will increase at a much slower rate of 1-3% per °C. The reason for the discrepancy between satellite observations of precipitation and models is not clear.

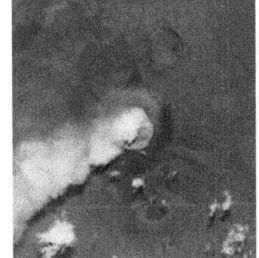

Identifying a Human "Fingerprint" Pattern in Atmospheric Water Vapor Changes.[23] Water vapor feedbacks are likely to play a key role in determining the magnitude of climate changes over the next century. Data from the satellite-based SSM/I show that the total atmospheric moisture content over oceans has increased by 0.41 kg m^{-2} per decade since 1988. Current climate models indicate that water vapor increases of this magnitude cannot be explained by natural variability alone. In a formal detection and attribution analysis using the pooled results from 22 different climate models, the simulated "fingerprint" pattern of anthropogenic changes in water vapor is identifiable with high statistical confidence in the SSM/I data. Experiments have been conducted in which forcing factors related to greenhouse gases, solar radiation, and volcanic eruptions are varied individually. The study found a decrease in water vapor levels due to cooling induced by massive volcanic eruptions (e.g., Pinatubo in 1991). The "fingerprint match" of the observed levels of water vapor indicates that the increases are primarily due to human-caused increases in greenhouse gases and not to solar or volcanic forcing. The findings provide evidence of an emerging anthropogenic signal in the moisture content of Earth's atmosphere.

Assessing the Causes of Global Stratospheric Temperature Changes.[24] The substantial cooling of the global lower stratosphere between 1979 and 2003 occurred in two pronounced step-like transitions. These arose in the aftermath of two major volcanic eruptions, with each cooling transition followed by a period of relatively steady temperatures. Climate model simulations demonstrate that the space-time structure of the observed cooling is largely attributable to the combined effect of changes in both anthropogenic factors (ozone depletion and increases in well-mixed greenhouse gases) and natural factors (solar irradiance variation and volcanic aerosols). The anthropogenic factors drove the overall cooling during the period, while the natural ones modulated the evolution of the cooling. The quantitative understanding of the global-mean lower stratospheric temperature evolution, including the distinct roles of the different forcing agents, is now at a high confidence level.

Black Carbon Impacts on the Asian Monsoon.[25] There is a growing awareness that black carbon aerosols, with their properties of both absorbing and reflecting solar radiation, may be contributing to significant climate change in the Indian monsoon region of south Asia. To address this problem, a six-member ensemble of 20th-century simulations with changes to only time-evolving global distributions of black carbon aerosols in a global coupled climate model was analyzed to study the effects of black carbon aerosols on the Indian monsoon. The black carbon aerosols increase lower tropospheric heating over south Asia and reduce the amount of solar radiation reaching the surface during the dry season. The increased north-south tropospheric temperature gradient in the pre-monsoon months of March, April, and May, particularly between the elevated heat source of the Tibetan Plateau and areas to the south, contributes to enhanced precipitation over India in those months. With the onset of the monsoon, the reduced surface temperatures in the Bay of Bengal, Arabian Sea, and over India that extend to the Himalayas act to reduce monsoon rainfall over India itself, with some small increases over the Tibetan Plateau. During the summer monsoon season, the model experiments show that black carbon aerosols have likely contributed to observed decreasing precipitation trends over parts of India, Bangladesh, Burma, and Thailand.

Potential Impacts of Future Temperature Changes on Tropical Glaciers.[26,27,28] Recent studies have considered the potential impact of future climate warming on tropical glaciers in the Americas, which provide a vital source of fresh water (as well as hydropower) for local communities. One study examines projected changes in mean annual free-air temperatures between 1990 to 1999 and 2090 to 2099 along a section of the American Cordillera from Alaska to Chile. The results show a general tendency toward increased warming at higher altitudes, with maximum temperature increases projected to occur in the high mountains of Peru, Bolivia, and northeast Chile. Such changes would have major consequences for mountain glaciers and for communities that depend on glacier meltwater for their water supplies. Because the Andean region has become highly dependent on hydropower (comprising over 50% of total energy supply in Ecuador and about 80% in Peru), rapid glacier retreat in the Andes also has substantial economic implications for the region.

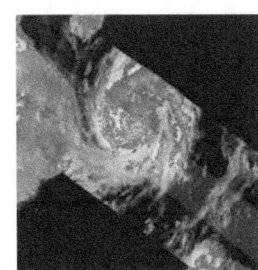

HIGHLIGHTS OF PLANS FOR FY 2009

Completing a Historical Reanalysis of the Atmosphere for the 20th Century. CCSP research has shown the feasibility of using modern data assimilation techniques together with observations of sea level pressure to produce a global analysis of tropospheric weather patterns at 6-hour temporal resolution for periods prior to the advent of extensive upper-air observations around 1950. Production of this first historical reanalysis will be completed in 2009. This reanalysis will provide, for the first time, a high-temporal resolution objective analysis of tropospheric weather and climate conditions over the entire 20th century. This historical reanalysis will help researchers address questions about the range of natural variability of high-impact events like floods, droughts, hurricanes, and extratropical cyclones, and how ENSO and other climate modes alter these events. This reanalysis will also help to clarify the origins of climate variations that produced major societal impacts and profoundly influenced policies, including the 1930s Dust Bowl drought and the prolonged cool, very wet period in the western United States early in the 20th century that led to over-allocation of Colorado River water through the 1922 Colorado Compact.

This activity will address Goals 1 and 3 and Questions 4.2, 4.4, and 4.5 of the CCSP Strategic Plan.

Developing a National Capacity for Integrated Earth System Analysis. CCSP continues to place high priority on developing a national capacity for integrated Earth system analysis that will enable scientists to better understand interactions among Earth system components that may produce rapid or unexpected climate changes, as well as high-impact weather and climate events. Achieving this capability requires parallel advancements in coupled Earth system modeling and the ability to assimilate Earth system observations into

models. A coupled ocean-atmosphere-land-ice model will serve as the basis for developing a reanalysis of the evolution of these Earth system components over the period of extensive satellite observations from 1979 to the present. This reanalysis will also help support research and modeling capabilities required to improve seasonal-to-interannual climate forecasts.

This activity will address Goals 1 and 3 and Questions 4.2, 4.4, and 4.5 of the CCSP Strategic Plan.

Advancing Understanding of the Causes of Drought. **Under the auspices of the U.S. Climate** Variability and Predictability (CLIVAR) project, a number of CCSP agencies initiated an activity in FY 2007 to support research into the physical and dynamical mechanisms leading to drought and the mechanisms through which drought may change as climate changes. The Drought in Coupled Models Project (DRICOMP) focuses on evaluating a variety of existing model products to address issues like the roles of the oceans and the seasonal cycle in drought, the impacts of drought on water availability, and distinctions between drought and drying. Based on a competition during FY 2007, 16 projects were selected to analyze and evaluate unforced control runs archived as part of the World Climate Research Programme (WCRP) Coupled Model Intercomparison Project 3 (CMIP3) multi-model data set, as well as the multi-model simulations of 20th-century climate and the long stabilized simulations with forcing held fixed at future climate conditions. The objective of DRICOMP is to increase community-wide diagnostic research into the physical mechanisms of drought and to evaluate its simulation in current models. DRICOMP will lead to more robust evaluations of model projections of drought risk and severity and thus to a better quantification of the uncertainty in such projections. In FY 2009, a workshop will be held during which results of the DRICOMP project will be summarized, and results from this interagency-supported project will be published.

This activity will address Goals 1 and 3
and Questions 4.2, 4.3, and 4.5 of the CCSP Strategic Plan.

Abrupt Climate Change and the Atlantic Meridional Overturning Circulation. The AMOC is thought to play an important role in the global ocean circulation, and therefore consequences to climate resulting from its variability, particularly in the North Atlantic, could be significant. Despite the potential for serious impacts, there are substantial gaps in knowledge concerning the AMOC and how it varies within the global climate system. Responding to recommendations in the report *Charting the Course for Ocean Science in the United States for the Next Decade:*

An Ocean Research Priorities Plan and Implementation Strategy by the National Science and Technology Council Joint Subcommittee on Ocean Science and Technology, several CCSP agencies will implement a program in FY 2009 that will lead to new capabilities for monitoring and predicting AMOC changes. In addition, it is anticipated that an abrupt climate change modeling activity initiated in FY 2008 will be continued in FY 2009. This activity would focus on examining attribution of past abrupt climate change, as well as the likelihood of future abrupt climate change based on climate change projections using dynamical coupled climate models.

This activity will address Goals 1 and 3 and Questions 4.1, 4.2, 4.3, and 4.5 of the CCSP Strategic Plan.

Improving High-End Modeling Capabilities for Decision Support. **Modeling efforts will** emphasize the continuing development of global and regional high-resolution atmospheric models. One major emphasis is to improve simulations of the climatology, interannual variability, and trends in Atlantic hurricane activity. These dynamical modeling studies promise to provide essential information to complement observational research based on historical hurricane records. In addition, through the North American Regional Climate Change Assessment Program (NARCCAP), several regional models will be used to downscale global model projections used in the IPCC Fourth Assessment Report to better estimate future climate changes over North America. NARCCAP modeling efforts will be complemented by global atmospheric model runs at high resolution (~50 km) over specific time periods performed at the Geophysical Fluid Dynamics Laboratory and the National Center for Atmospheric Research.

This activity will address CCSP Goal 3 and Questions 4.1, 4.2, 4.4, and 4.5 of the CCSP Strategic Plan.

CLIMATE VARIABILITY AND CHANGE
CHAPTER REFERENCES

1) **Donnelly**, J.P. and J.D. Woodruff, 2007: Intense hurricane activity over the past 5000 years controlled by El Niño and the West African monsoon. *Nature*, **447**, 465-468.

2) **Nyberg**, J., B.A. Malmgren, A. Winter, M.R. Jury, K.H. Kilbourne, and T.M. Quinn, 2007: Low Atlantic hurricane activity in the 1970s and 1980s compared to the past 270 years. *Nature*, **447**, 698-702.

3) **Mann**, M.E. and K.A. Emanuel, 2006: Atlantic hurricane trends linked to climate change. *Eos, Transactions of the American Geophysical Union*, **87**, 233, 238, 241.

4) **Landsea**, C.W., 2007: Counting Atlantic tropical cyclones back to 1900. *Eos, Transactions of the American Geophysical Union*, **88**, 197, 202.

5) **Mann**, M.E., K.A. Emanuel, G.J. Holland, and P.J. Webster, 2007: Atlantic tropical cyclones revisited. *Eos, Transactions of the American Geophysical Union*, **88**, 349-350.

6) **Shepherd**, J.M. and T. Knutson, 2007: The current debate on the linkage between global warming and hurricanes. *Geography Compass*, **1**, 1-24.

7) **Vecchi**, G.A. and B.J. Soden, 2007: Increased tropical Atlantic wind shear in model projections of global warming. *Geophysical Research Letters*, **34**, 1-5, doi:10.1029/2006GL028905.

8) **Knutson**, T.R., J.J. Sirutis, S.T. Garner, I.M. Held, and R.E. Tuleya, 2007: Simulation of the recent multi-decadal increase of Atlantic hurricane activity using an 18-km grid regional model. *Bulletin of the American Meteorological Society*, **88**, 1549-1565.

9) **Seager**, R., M. Ting, I. Held, Y. Kushnir, J. Lu, G. Vecchi, H.-P. Huang, N. Harnik, A. Leetmaa, N.-C. Lau, C. Li, J. Velez, and N. Naik, 2007: Model projections of an imminent transition to a more arid climate in southwestern North America. *Science*, **316**, 1181-1184.

10) **Stan**, C. and B.P. Kirtman, 2008: The influence of atmospheric noise and uncertainty in ocean initial conditions on the limit of predictability in a coupled GCM. *Journal of Climate*, **21**, doi:10.1175/2007JCLI2071.1.

11) **Huang**, B., Z.-Z. Hu, and B. Jha, 2007: Evolution of model systematic errors in the tropical Atlantic basin from coupled climate hindcasts. *Climate Dynamics*, **28**, 661-682.

12) **Hu**, Z.-Z. and B. Huang, 2007: The predictive skill and the most predictable pattern in the tropical Atlantic: The effect of ENSO. *Monthly Weather Review*, **135**, 1786-1806.

13) **Cunningham**, S.A., T. Kanzow, D. Rayner, M.O. Baringer, W.E. Johns, J. Marotzke, H.R. Longworth, E.M. Grant, J.J.-M. Hirschi, L.M. Beal, C.S. Meinen, and H.L. Bryden, 2007: Temporal variability of the Atlantic meridional overturning circulation at 26.5° N. *Science*, **317**, 935-938.

14) **Kanzow**, T., S. Cunningham, D. Rayner, J.J.-M. Hirschi, W.E. Johns, M.O. Baringer, H.L. Bryden, L.M. Beal, C.S. Meinen, and J. Marotzke, 2007: Observed flow compensation associated with the MOC at 26.5°N in the Atlantic. *Science*, **317**, 938-941.

15) **Meier**, M.F., M.B. Dyurgerov, U.K. Rick, S. O'Neel, W.T. Pfeffer, R.S. Anderson, S.P. Anderson, and A.F. Glazovsky, 2007: Glaciers dominate eustatic sea-level rise in the 21st century. *Science*, **317**, 1064-1066.

16) **Tedesco**, M., 2007: Snowmelt detection over the Greenland ice sheet from SSM/I brightness temperature daily variations. *Geophysical Research Letters*, **34**, L02504, doi:1029/2006GL028466.

17) **Howat**, I.M., I. Joughin, and T.A. Scambos, 2007: Rapid changes in ice discharge from Greenland outlet glaciers. *Science*, **315**, 1559-1561.

18) **Stroeve**, J.M., M. Holland, W. Meier, T. Scambos, and M. Serreze, 2007: Arctic sea ice decline: faster than forecast. *Geophysical Research Letters*, **34**, L09501, doi:10.1029/2007GL029703.

19) **Francis**, J.A. and E. Hunter, 2007: Drivers of declining sea ice in the Arctic winter: A tale of two seas. *Geophysical Research Letters*, **34**, L17503, doi:10:1029/2007GL030995.

20) **Zhang**, X., F.W. Zwiers, G.C. Hegerl, H.F. Lambert, N.P. Gillett, S. Solomon, P.A. Stott, and T. Nozawa, 2007: Detection of human influence on twentieth-century precipitation trends. *Nature*, **448**, 461-466.

21) **Wentz**, F.J., L. Ricciardulli, K. Hilburn, and C. Mears, 2007: How much more rain will global warming bring. *Science*, **317**, 233-235.

CLIMATE VARIABILITY AND CHANGE
CHAPTER REFERENCES (CONTINUED)

22) **Lorenz**, D.J., and E.T. DeWeaver, 2007: The response of the extratropical hydrological cycle to global warming. *Journal of Climate*, **20**, 3470-3484.

23) **Santer**, B.D., C. Mears, F.J. Wentz, K.E. Taylor, P.J. Gleckler, T.M.L. Wigley, T.P. Barnett, J.S. Boyle, W. Brüggemann, N.P. Gillett, S.A. Klein, G.A. Meehl, T. Nozawa, D.W. Pierce, P.A. Stott, W.M. Washington, and M.F. Wehner, 2007: Identification of human-induced changes in atmospheric moisture content. *Proceedings of the National Academy of Sciences*, **104(39)**, 15248-15253.

24) **Ramaswamy**, V., M.D. Schwarzkopf, W.J. Randel, B.D. Santer, B.J. Soden, and G.L. Stenchikov, 2006: Anthropogenic and natural influences in the evolution of lower stratospheric cooling. *Science*, **311**, 1138-1141.

25) **Meehl**, G.A., J.M. Arblaster, and W.D. Collins, 2008: Effects of black carbon aerosols on the Indian monsoon. *Journal of Climate*, **21**, 2869-2882.

26) **Bradley**, R.S., M. Vuille, H.F. Diaz, and W. Vergara, 2006: Threats to water supplies in the tropical Andes. *Science*, **312**, 1755-1756.

27) **Thompson**, L.G., E. Mosley-Thompson, H. Brecher, M. Davis, B. León, D. Les, P-Nan Lin, T. Mashiotta, and K. Mountain, 2006: Abrupt tropical climate change: past and present. *Proceedings of the National Academy of Sciences*, **103**, 10536-10543.

28) **Vergara**, W., A. Deeb, A. Valencia, R.S. Bradley, B. Francou, A. Zarzar, A. Grunwaldt, and S. Haeussling, 2007: The economic impacts of rapid glacier retreat in the Andes. *Eos, Transactions of the American Geophysical Union*, **88**, 261, 264.

3 | Global Water Cycle

Strategic Research Questions

5.1 What are the mechanisms and processes responsible for the maintenance and variability of the water cycle; are the characteristics of the cycle changing and, if so, to what extent are human activities responsible for those changes?

5.2 How do feedback processes control the interactions between the global water cycle and other parts of the climate system (e.g., carbon cycle, energy), and how are these feedbacks changing over time?

5.3 What are the key uncertainties in seasonal to interannual predictions and long-term projections of water cycle variables, and what improvements are needed in global and regional models to reduce these uncertainties?

5.4 What are the consequences over a range of space and time scales of water cycle variability and change for human societies and ecosystems, and how do they interact with the Earth system to affect sediment transport and nutrient and biogeochemical cycles?

5.5 How can global water cycle information be used to inform decision processes in the context of changing water resource conditions and policies?

See Chapter 5 of the *Strategic Plan for the U.S. Climate Change Science Program* for detailed discussion of these research questions.

The global water (and energy) cycle plays a critical role in the functioning of the Earth system. Through complex interactions, the global water cycle integrates the physical, chemical, and biological processes that sustain ecosystems and influence climate and related global change. Inadequate understanding of the water/energy cycle is one of the key sources of uncertainty in climate prediction and climate change projections. Clouds, precipitation, and water vapor play important roles in feedbacks that are not well represented in many climate models. These processes alter surface and atmospheric heating and cooling rates, leading to adjustments in atmospheric circulation and precipitation patterns. Improved understanding of these processes will be essential to developing options for responding to the consequences of water cycle variability and change. For assessing the impacts of global and regional climate change on human societies, industrial and economic systems, and natural and managed ecosystems, water is considered a more rigid or critical constraint or limiting factor than temperature. To address these issues the CCSP Global Water Cycle element expends considerable effort to improve observations, data assimilation, and modeling/prediction systems that in turn deliver the information necessary for decision-support tools and assessments that provide a basis for "best practices" in the management of water resources.

The ultimate goal of water cycle research is to provide a solid foundation for decisions and investments by policymakers, managers, and individuals, be it at the Federal, State, or local level. Achieving this goal requires a program of activities that significantly improves understanding of water/energy cycle processes, incorporates this understanding in an integrated modeling/prediction framework, and tests predictions and data products in real decisionmaking contexts. In order to demonstrate techniques and effectiveness to potential users, the Global Water Cycle program also aims to expedite the transfer of science results from the research/experimental realm to operational applications.

Significant progress is being made in the understanding of cloud properties, the direct and indirect effect of aerosols on cloud and precipitation processes, and the interaction of cloud systems with land surface hydrological conditions. To this end, the Global Water Cycle program's first integrated priority activity was completed in June 2007 through multi-agency participation in DOE's Cloud and Land Surface Interaction Campaign (CLASIC). Comprehensive satellite monitoring of water cycle parameters— such as global precipitation and cloud structure in storm systems and hurricanes [with the Tropical Rainfall Measuring Mission (TRMM) and Cloudsat], soil moisture [with the Special Sensor Microwave/Imager (SSM/I) and the Advanced Microwave Scanning Radiometer-Earth Observing System (AMSR-E)], and water bodies [with the Gravity Recovery and Climate Experiment (GRACE)], as well as more accurate atmospheric

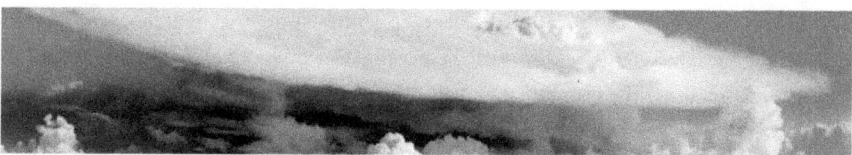

profiles of temperature, humidity, and land/ocean surface parameters (with the multi-sensor Terra and Aqua satellites)—have resulted in integrated data sets and improved models of the Earth system. Ensemble Kalman filtering techniques have demonstrated the potential use of satellite-derived surface soil moisture estimates to improve the characterization of soil moisture at depth in land information systems. The incorporation of research results in models has led to better simulation/prediction capabilities for hydroclimatic variables. Multi-model and ensemble modeling techniques developed by CCSP have led to improved seasonal predictions of both the atmospheric and terrestrial hydrological cycle. Techniques have also been developed by USDA's Agricultural Research Service, DOI/USGS, and the DOI Bureau of Reclamation, in collaboration with NOAA, NASA, EPA, and DOE, among others, for the downscaling of intra-seasonal and seasonal precipitation forecasts to temporal scales consistent with the input requirements for agricultural and water resources management as well as conservation planning and decision-support tools. Experimental seasonal hydrological prediction systems have been developed that use multi-model climate forecast products and empirical tools to "force" land/hydrological prediction models.

HIGHLIGHTS OF RECENT RESEARCH

The following are selected highlights of recent research supported by the CCSP-participating agencies. These research results address the strategic research questions on the global water cycle identified in the CCSP Strategic Plan.

Interagency Cloud and Land Surface Interaction Field Experiment. **Improved** understanding of the water/energy cycle is a key factor in reducing uncertainty in climate prediction. Parameters such as regional scale soil moisture and key processes involving the interactions between cloud formation and the moisture availability of land surfaces are characteristic of needed improvements. The development of continental cumulus convection is strongly modulated by land surface conditions, while influencing the land surface through rain-induced changes in soil moisture and photosynthesis. To improve understanding of cloud properties, the direct and indirect effect of aerosols on cloud formation processes, and interactions between clouds and the land surface, the first of a series of interagency CLASIC field studies was conducted in June 2007 (see <science.arm.gov/clasic>. The region surrounding DOE's Southern Great Plains

(SGP) site in Oklahoma was chosen for the field experiment due to its extensive surface-based instrumented facilities. In addition, three "supersites" were also heavily instrumented to obtain ground-based measurements to link observed carbon and moisture fluxes to atmospheric structure. Several instrumented research aircraft were provided by the CCSP agencies involved. Collaborations were established between CLASIC and the North American Carbon Program's Mid-Continent Intensive (MCI) study, recognizing the strong synergy between measurements in SGP and the northern MCI locations, particularly because of air masses flowing from south to north and the influence of the land surface on atmospheric concentrations of aerosols, gases, and other constituents (see Figure 8).

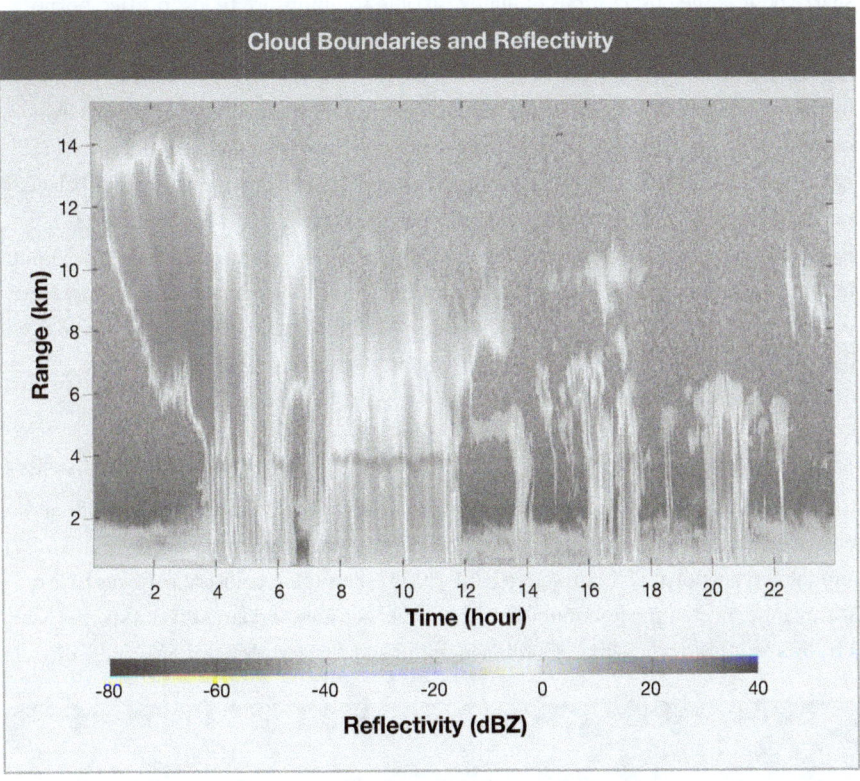

Figure 8: Cloud Boundaries and Reflectivity. As part of the Cloud and Land Surface Interaction Campaign (CLASIC), three supersites were instrumented to obtain ground-based measurements to link observed carbon fluxes to atmospheric structure. Nine aircraft—including a helicopter—participated. The SGP site's Central Facility served as the primary source of information for cloud distribution and carbon feedbacks. The other two supersites were located in pastured lands near the Little Washita Watershed and oak forests near Okmulgee State Park. This image is from a millimeter wavelength cloud radar, which probes the extent and composition of clouds to provide information about cloud boundaries and reflectivity. On the morning of June 14, the radar detected a thunderstorm as it descended on the SGP site, followed by about 5 hours of heavy rain and then brief showers throughout the rest of the day. These data represent a rather complicated case from a modeling perspective, and therefore the need to better understand interactions and feedbacks at the land surface. *Credit: W. Ferrell, DOE.*

Improved Methodology to Validate Remotely Sensed Soil Moisture Products.[1] A novel methodology has been developed to validate the added value of remotely sensed soil moisture products [e.g., from AMSR-E, TRMM microwave imager (TMI)] using a Kalman filter-based strategy that does not require the availability of ground-based soil moisture measurements, and one that can be applied anywhere relatively high-quality rain gauge observations are available (e.g., the contiguous United States). The validation of global remote sensing products is typically based on the use of test bed sites in data-rich areas to characterize retrieval accuracy and value in higher order applications. However, the ability to validate space-borne soil moisture products against ground-based observations is currently limited by difficulties in maintaining soil moisture instrument networks and upscaling sparse point-scale observations of highly variable soil moisture fields to space-borne footprint scales (10-30 km). The new approach provides a quantitative measure of soil moisture with an accuracy linked to that of currently available global rainfall products by means of a simple water balance model. Results indicate that even retrievals from non-optimal X-band frequency sensors over heavily vegetated areas significantly enhanced the quality of soil moisture predictions derived from a simple water balance model and the space-borne precipitation data set. Overall, this study represents an important benchmark that remotely sensed soil moisture products must improve upon in order to contribute value to global land surface modeling applications. The presence of detectable skill at X-band frequencies bodes well for future space-borne missions based on lower frequency L-band measurements better suited for soil moisture measurements and penetration through dense vegetation canopies.

Influence of Land Cover and Soil Moisture on Heat Fluxes.[2] Analysis of aircraft, surface-flux tower, and radar wind profiler data on six fair weather days with southerly winds and near-clear skies from the Cooperative Atmosphere-Surface Exchange Study (CASES-97) and the International H_2O Project (IHOP-2002) shows that land-use patterns have a strong influence on the horizontal distribution of sensible and latent heat fluxes (H and LE) over southeastern Kansas. Combined with Land Surface Model (LSM) runs, the

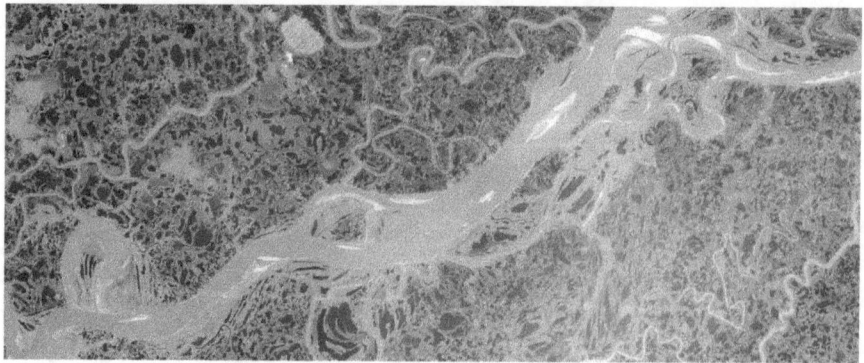

data suggest that soil moisture influences the relative magnitude of H and LE horizontal variation. In both field programs, H maxima occurred over dormant/sparse vegetation, with a minimum over green vegetation. To a lesser degree, LE maxima occurred over green vegetation, with a minimum over dormant/sparse vegetation. Small day-to-day differences in flux distribution occurred due to the effect of wind direction and speed and surface buoyancy fluxes at the scale of the surface heterogeneity as well as statistical uncertainty. The soil moisture and length of time after rainfall affect the amplitude and coherence of the LE and H horizontal patterns. Terrain could also modulate the horizontal variability in fluxes in this region.

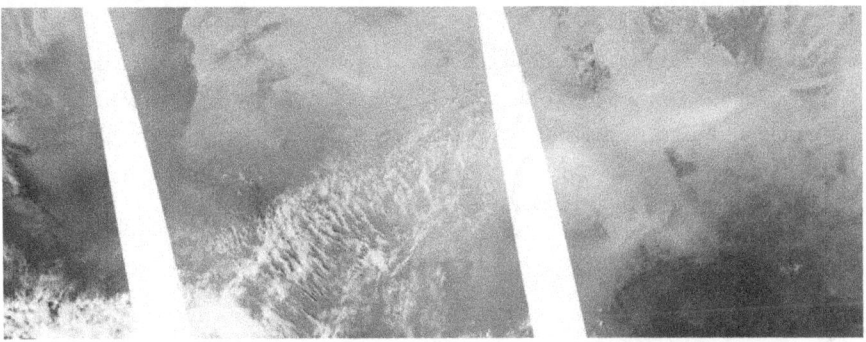

Impact of Desert Dust Radiative Forcing on Sahel Precipitation.[3] A recent investigation considered the role of radiative forcing by dust particles in the Sahelian drought, which occurred over the last 3 decades of the 20th century. The study compared atmospheric general circulation model simulations with meteorological and hydrological measurements. In comparison to previous studies, dust particles that are less absorptive of solar radiation and more emissive at long wavelengths were used in the present study. Cooling of the atmosphere due to dust radiative forcing was found to play an important role in reducing the precipitation over North Africa. The newly modeled circulation responses to this cooling over North Africa provide better agreement with the observations made in dry years in the Sahel region. The results are important because they show that the direct radiative forcing by dust has played a role in the observed droughts in the Sahel comparable to the roles played by sea surface temperatures and vegetation, which have been extensively studied. These results also provide a mechanism whereby drought in the Sahel region can cause increased dust, which then feeds back to cause a further precipitation reduction.

Climate Variability and Fluctuations in Daily Precipitation over the United States.[4] Fluctuations in the frequency and intensity of daily precipitation over the United States during the period 1948 to 2004 were identified and linked to leading sources of interannual and interdecadal climate variability. The El Niño-Southern Oscillation (ENSO) phenomenon

was implicated in interannual fluctuations while the Pacific Decadal Oscillation (PDO) and the Arctic Oscillation (AO) were linked to recent interdecadal fluctuations. For the conterminous United States as a whole, there have been increases in the annual frequency of occurrence of wet days and heavy precipitation days and in the mean daily and annual total precipitation over the past several decades, though these changes have not been uniform. The study explored the possibility of significant natural forcing of these interdecadal variations in precipitation, and found that the PDO is associated with these fluctuations over the western and southern United States, while the AO is also associated with them but to a much lesser extent over the southeastern United States. Because the interdecadal fluctuations are linked to changes in the global-scale circulation and sea surface temperatures associated with the PDO, the results imply that a significant portion of the skill of climate models in anticipating fluctuations in daily precipitation statistics over the United States will arise from an ability to forecast the temporal and spatial variability of the interdecadal shifts in tropical precipitation and in the associated teleconnection patterns into the mid-latitudes.

Consensus U.S. Drought Monitor. The National Drought Mitigation Center (NDMC) has developed an "integrated" Drought Monitor—a synthesis of multiple indices, outlooks, and new accounts that represents a consensus of Federal and academic scientists (see <drought.unl.edu/dm/monitor.html>). The experimental drought monitor product shown in Figure 9, to be refined over time, has improved techniques that are found to better reflect the needs of decisionmakers and others who use the information. The Drought Monitor integrates information from a range of data on rainfall, snowpack, streamflow, and other water supply indicators into a comprehensible picture. Drought measures used include Percent of Normal; Standardized Precipitation Index; Palmer Drought Severity Index; Crop Moisture Index; Surface Water Supply Index; Reclamation Drought Index; and Deciles, among others. With an emphasis on preparation and risk management, rather than crisis management, NDMC helps people and institutions develop and implement measures to reduce vulnerabilities to drought. The NDMC works in partnership with several of the CCSP agencies. These activities also contribute to the National Integrated Drought Information System (NIDIS).

Regional Climate Model for North and South America.[5,6] Proper evaluation of climate change at regional scales is crucial for society planning to mitigate the impact of global climate change. A study using the National Centers for Environmental Prediction Eta regional climate model indicates that under proper choices of model domain size, imposed lateral boundary conditions, and horizontal resolution, the regional model is capable of producing better regional features, such as precipitation and the low-level jet than the general circulation model that provides the lateral boundary conditions to the regional model.

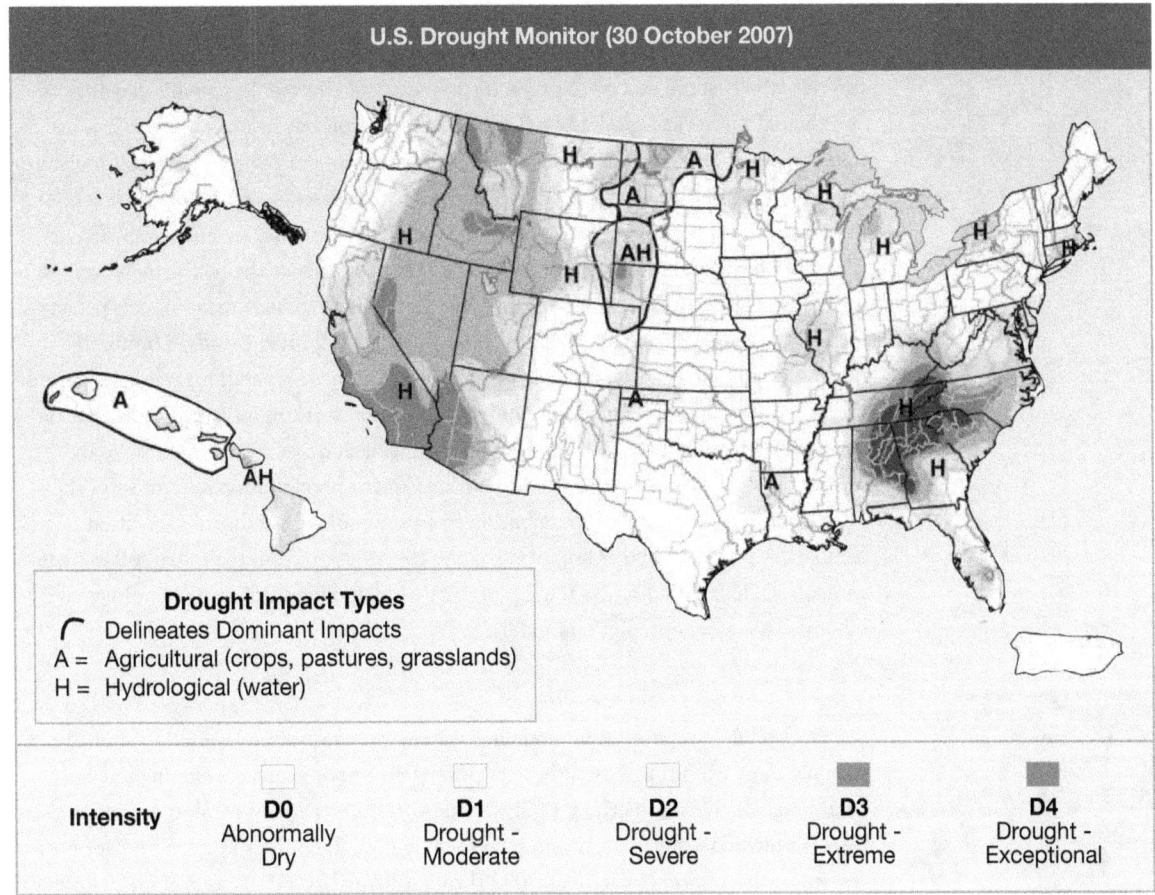

Figure 9: U.S. Drought Monitor (30 October 2007). The U.S. Drought Monitor blends quantitative measures of drought and experts' best judgment into a single map every week. The Drought Monitor grew out of a Western Governors' Association initiative to provide timely and understandable scientific information on water supply and drought for policymakers. The U.S. Drought Monitor—a Federal, State, and academic partnership—has been operational since 1999, and is produced by a rotating group of authors from USDA, NOAA, and NDMC. It incorporates review from a group of more than 250 climatologists, hydrologists, meteorologists, extension agents, and others across the nation. Each week an author revises the previous map based on rain, snow and other events, observers' reports of how drought is affecting crops and wildlife, and other indicators. Authors balance conflicting data and reports to come up with a new map every Wednesday afternoon, with release the following Thursday morning. *Credit: USDA, NOAA, and NDMC (University of Nebraska-Lincoln).*

Impact of Vegetation and Soil Moisture Feedback on Precipitation.[7,8] Accurate seasonal climate predictions of precipitation are critical for agriculture, water management and planning, and for the mitigation of natural hazards. Both large-scale oceanic forcing and local land surface conditions are important factors in determining precipitation over the United States. Past seasonal predictions have primarily relied on sea surface temperature due to its slow variation. Research over the past decade has produced abundant evidence that positive feedback between soil moisture and precipitation over most of the United States promotes the persistence of seasonal hydrological conditions. The time scale of

this persistence, or the memory length of the land-atmosphere system, can be as long as 3 months in late spring and summer, suggesting that the slowly varying soil moisture can potentially serve as a good predictor for seasonal climate. To use soil moisture as a "predictor," numerical model-based seasonal prediction requires accurate soil moisture initialization and the realistic simulation of important processes involved in soil moisture-precipitation coupling. One of these processes is the seasonal vegetation feedback. At the seasonal time scale, vegetation responds to concurrent and cumulative hydro-meteorological conditions and feeds back to further influence the hydro-meteorological processes. This feedback has largely been neglected in the past as most models prescribe, instead of predict, the seasonal course of vegetation. In a recent study, a predictive phenology scheme (which predicts the seasonal variation of vegetation) was incorporated into a coupled land-atmosphere model, and the impact of soil moisture anomalies on subsequent precipitation examined. Vegetation feedback was found to enhance the impact of wet soil moisture anomalies on subsequent precipitation over most of the Mississippi River Basin. The contribution from soil moisture-induced vegetation feedback was found to be as important as the contribution from the initial soil moisture anomalies. This finding shows the importance of including predictive phenology schemes in seasonal prediction models.

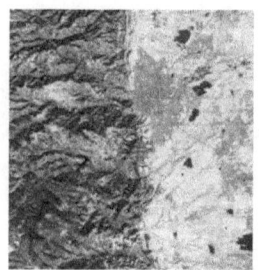

A High-Resolution Meteorological Distribution Model for Atmospheric, Hydrologic, and Ecologic Applications.[9] A snow evolution modeling system (SnowModel) was used to simulate seasonal snow evolution across three 30-km by 30-km domains enveloping the Cold Land Processes Field Experiment (CLPX) meso-cell study areas in Colorado. Simulations were performed using a 30-m grid increment and spanned the snow accumulation season for this region (1 October 2002 through 1 April 2003). Meteorological forcing was provided by 27 meteorological stations and 75 atmospheric analysis grid points distributed across the model simulation domains using a micrometeorological distribution model (MicroMet). The simulations included a data assimilation sub-model (SnowAssim) that adjusted snow water equivalent (SWE) toward a collection of ground-based and airborne SWE observations. Simulated SWE distributions displayed considerably more spatial heterogeneity compared with observations alone, and the simulated distribution patterns closely fit understanding of snow evolution processes and observed snow depths.

Land Surface-Atmosphere Interactions studied by Comparing Simulated vs. Observed Fluxes and Feedbacks.[10] Land atmosphere interactions in the Weather Research and Forecasting (WRF) model and the Noah Land Surface Model (Noah LSM) were analyzed by comparing simulated fluxes and feedbacks to *in situ* and remotely sensed observations. Vegetation cover, vegetation water content, and land surface temperature data acquired from remotely sensed platforms are strongly correlated in semiarid regions, such as the

North American Monsoon Region, compared to more humid regions. The assimilation of these data, including their covariance, into the Noah LSM improved the simulation of soil moisture and other land surface fields. However, the latent heat flux to the atmosphere, and therefore likelihood of precipitation, were found to be overestimated due to the parameterization of vegetation physiology.

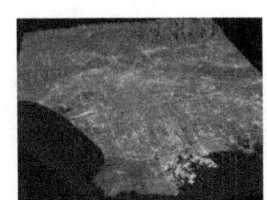

New Water Stress Index to Assess the Impacts of Environmental Change on Water Availability and Use in the United States.[11] **Watershed** water stress is caused by both water availability (i.e., lack of supply) and use (i.e., demand), both of which are influenced by ecosystem conditions and humans. This study developed a water stress index that integrates both natural and anthropogenic effects on water availability and use. Future availability and use scenarios were modeled via changes in climate, land management, land use/land cover, and population. Results suggest that population growth will greatly increase water use in metropolitan areas, but overall, changes in population will have little impact on total water demand over the next 40 years. In contrast, changes in air temperature and precipitation will likely affect regional water availability significantly in coming decades.

Will Thunderstorms be Stronger in a Warmer Climate?[12] **How** thunderstorm intensity will change with global climate change is an interesting and particularly important question. While there is some evidence that the intensity of hurricanes will increase, predictions for the nature of thunderstorms that are not part of hurricanes is lacking. Using a global climate model with a new parameterization of vertical velocities (or updrafts) in thunderstorms, Atmospheric Radiation Measurement (ARM) researchers examined how

the intensity of thunderstorms varies from region to region over ocean and land and how the vertical velocity will increase in a warmer climate. Their results indicate that a simple estimate of the upward velocity of thunderstorm updrafts in a global climate model reproduces observed land-ocean differences in thunderstorm intensity. Under a climate change scenario, updrafts strengthen by about 1 m s^{-1} in the lightning-producing regions of continental thunderstorms, primarily due to an upward shift in the freezing level. For the western United States, drying in the warmer climate reduces the frequency of thunderstorms that initiate forest fires, but the strongest storms occur 26% more often. For the central-eastern United States, stronger updrafts combined with weaker change of winds with height (or wind shear) suggest little change in severe storm occurrence with global climate change, but the most severe storms occur more often.

Rock Glaciers as Hydrologic Refugia in a Warming World.[11] Rock glaciers are widespread but little studied landforms in semi-arid mountain ranges of the world. Rock mantling insulates ice from solar heating, creating a lag in response to climate relative to typical glaciers and winter snowpacks. A classification system has been developed and used to survey over 400 features in canyons of the Sierra Nevada, California. It was found that these features are undocumented sources of mountain water, and provide wetland refugia for mountain biodiversity. As snowpacks diminish in the future, they will likely gain in local importance.

HIGHLIGHTS OF PLANS FOR FY 2009

The Global Water Cycle research element continues to pursue important long-term priorities. For example, insights into the formation and behavior of clouds and precipitation, including better characterizations of the phase changes of water in clouds and the phases and onset of precipitation, are emerging from field campaigns and model studies and will be promoted in continuing activities. Water vapor and cloud-radiation feedback are considered a critical part of global water cycle studies that need

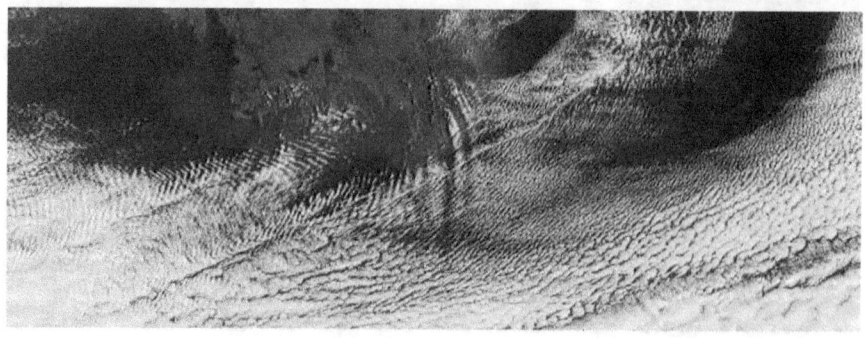

to be addressed to reduce uncertainties associated with climate change projections. The predictability of regional precipitation is another topic of vital interest: It will be assessed and better understood by ongoing diagnostic and modeling studies that identify the connections between regional- and global-scale phenomena, land surface conditions such as soil moisture and water table fluctuations, and the interface fluxes of energy and heat between the atmosphere and the land surface-vegetation-hydrology combination. Preliminary analyses from recent studies show promise of leading to earlier (and more accurate) predictions, improved ability to assess hazards and risks of extremes such as floods and droughts, and more efficient water resource management.

In FY 2009, continuing U.S. and global observations, field campaigns and experiments, improvements to data integration and analysis systems, diagnostic and predictive model development, and applications to decision-support systems will be priorities under the CCSP Global Water Cycle program. A fundamental objective is to ensure that observational capability is enhanced and improved, and that the data assimilation and modeling/prediction systems are more reliable and accurate at the point of application. Several promising results from the past years of research will be further explored with an aim to transfer this research knowledge to operational applications that provide societal benefit. To this end, the program has developed a series of priority activities for the FY 2009 (and beyond) time frame. Concurrently, a cohesive research strategy will be implemented to improve current deficiencies in understanding all aspects of the regional and global water cycle. Several science questions remain to be answered related to warnings of natural hazards and to the impact of global climate change, be it from natural or anthropogenic causes. A considerable research focus will be placed in FY 2009 and beyond on "downscaling" from the recently released IPCC Fourth Assessment Report climate projection model assessments. That is, downscaling, using a combination of high-resolution dynamical/statistical models, to the regional scales of critical import to decisionmaking as regards water resource management and associated infrastructures.

The program outlined for FY 2009 will lead to improvements in fundamental research, as well as in the planning and decisionmaking for, and management of, natural and human-made resources—a major aim of the program in addition to its fundamental research goals. A strong effort will continue to focus on major unresolved research issues that will require longer term commitments. To address both research and multi-sectoral applications needs, several initiatives will be launched in FY 2009 and beyond.

New Land Cover Visualization Tool for the Analysis of Land Cover Change with Time. **Land-cover** data has been a largely untapped information resource. Both natural processes and human influences shape land cover—the pattern of natural vegetation, agriculture, and

urban areas. Information about land cover is needed by managers of public and private lands, urban planners, agricultural experts, and scientists for studying such issues as climate change or invasive species. With increasing population and the challenging prospect of climate change, comprehensive information about the condition of land, and how it is changing, becomes more and more vital. Recently, USGS announced the launch of the new USGS Land Cover Visualization and Analysis Tool, which allows users to analyze, in specific detail, how land cover has changed over time. Designed for both novice and expert users, the web-based system provides an intuitive interface able to selectively view and analyze land-cover data from any web browser. USGS is soliciting users to evaluate the preview release of the application (see <lcat.usgs.gov>).

This activity will address CCSP Goals 1, 2, 3, 4, and 5 and Questions 5.1, 5.2, and 5.3 of the CCSP Strategic Plan.

Planning for an L-Band Soil Moisture Active-Passive (SMAP) Space-Based/Satellite Instrument Platform. Accurately measuring global terrestrial moisture is considered a high priority for a vast range of terrestrial hydrological science and applications that are of primary relevance to societal benefits. Following the prioritized recommendations of the National Research Council's Decadal Survey, CCSP-supported actions have been taken to begin development of the space-based instrument sensors required to measure terrestrial soil moisture with substantially improved accuracy compared to that currently provided by various passive microwave satellite instruments such as the SSM/I series and the microwave imager onboard the TRMM satellite (TRMM-TMI). NASA initiated planning with a dedicated SMAP design workshop in July 2007. Several CCSP agencies are involved in this process, as well as a diverse group of university researchers including representatives from the Massachusetts Institute of Technology who were involved with the Hydrosphere State Mission (Hydros) platform a few years ago. In a sense, Hydros provides the legacy for SMAP planning. Hydros was designated as an alternate to Aquarius but not finally selected (at that time) due to budgetary constraints. SMAP will contain an instrument suite broader in scope than Hydros. In FY 2009 and subsequent years, the SMAP mission will be designed, built, and launched contingent on budgetary considerations and national policy directives.

This activity will address CCSP Goals 1, 2, 3, 4, and 5
and Questions 5.1, 5.2, and 5.3 of the CCSP Strategic Plan.

Planning for a National Groundwater Recharge Monitoring Network. **In the United States and** around the world, groundwater is being withdrawn at unprecedented rates. Increasing pressures are being placed on hydrological systems to support urban expansion and rising demand from the agricultural industry, including the rapidly increasing demand arising from the bio-fuel industry as reported in many journals. Global and regional climate change imposes various scenarios and constraints regarding the sustainability of current water management infrastructures. To understand this situation better, planning has been initiated, led by USGS but with multi-agency collaboration, to develop a groundwater recharge monitoring network. The observation/measurement problem is complex and requires several research and monitoring aspects that currently do not exist. The Global Water Cycle interagency working group intends to integrate this key parameter (groundwater recharge) into the construct of an integrated, interagency project to address the "closure" of the water budget on the scales of a river basin and/or catchment area. Results from this research/observations exercise will contribute to development of improved models, which then can be applied to better assess water availability issues under a climate change scenario.

This activity will address CCSP Goals 1, 2, 3, 4, and 5
and Questions 5.1, 5.2, and 5.3 of the CCSP Strategic Plan.

International Hydrologic Ensemble Prediction Experiment. **CCSP's Climate Prediction** Program for the Americas (CPPA) will continue to support the Hydrologic Ensemble Prediction Experiment (HEPEX), an international project to demonstrate how to produce reliable hydrological predictions that can be used with confidence for decisionmaking. An initial experimental operational prototype system, using weather and climate forecasts to produce input forcing for hydrologic forecast models, will be tested at several National Weather Service (NWS) River Forecast Centers. This includes a strategy for seamless integration of weather and climate forecasts for all lead times from 1 hour to more than 1 year. Future CPPA plans include:

- Development of "supporting data sets" for hydrologic ensemble prediction research within the HEPEX test-beds and improved pre-processing methodologies
- Collaborative activities among the international science community, CPPA researchers, NWS , and Hydrologic Development and River Forecast Centers to improve seasonal hydrologic forecasting techniques
- Evaluation of improved models and techniques for making probabilistic hydrologic forecasts that integrate with CPPA research in land memory processes, orographic

processes, remote sensing, and climate predictions for possible operational application in NWS hydrologic forecast systems

- Demonstration of improvements in end-to-end hydroclimatic forecasting technologies at intra-seasonal and seasonal time scales.

This activity will address CCSP Goals 2, 3, and 5
and Questions 5.1, 5.2, and 5.3 of the CCSP Strategic Plan.

New "Holistic" Earth Surface Observations to Focus on the Science of Watershed Evolution. NSF has selected sites for three critical zone observatories (CZO) as an initial impetus to a long-term program. In FY 2009 and beyond, these observatories are designed to provide scientists with an understanding of what is called a "critical zone"—the region between the top of the forest canopy and the base of unweathered rock (the living environment)—and its response to climate and land-use changes. CZOs represent the first set of systems-based observatories dedicated to Earth surface processes. Scientists at each CZO will investigate the integration and coupling of Earth surface processes and how they are affected by the presence and flux of freshwater. CZOs will use field and analytical research methods, space-based remote sensing, and theoretical techniques. These projects will add to the environmental sensor networks already in place and those planned by NSF, including EarthScope, the National Ecological Observatory Network, and the Ocean Observatories Network. The CCSP Global Water Cycle interagency process will explore means by which sites such as CZOs and others can be complemented with additional multi-agency observational capabilities to meet the broader science objectives of the research element and, in particular, the planned Global Water Cycle priority activities/projects.

This activity will address CCSP Goals 1, 2, 3, and 4
and Questions 5.1, 5.2, and 5.3 of the CCSP Strategic Plan.

GLOBAL WATER CYCLE
CHAPTER REFERENCES

1) **Crow**, T.W, 2007: A novel method for quantifying value in spaceborne soil moisture retrievals. *Journal of Hydrometeorology*, **8**, 56-66, doi:10.1175/JHM533.1.

2) **Lemone**, M.A., F. Chen, J.G. Alfieri, M. Tewari, B. Geerts, Q. Miao, R.L. Grossman, and R.L. Coulter, 2007: Influence of land cover and soil moisture on the horizontal distribution of sensible and latent heat fluxes in Southeast Kansas during IHOP-2002 and CASES-97. *Journal of Hydrometeorology*, **8**, 68-87.

3) **Yoshioka**, M., N.M. Mahowald, A.J. Conley, W.D. Collins, D.F. Fillmore, C.S. Zender, and D.B. Coleman, 2007: Impact of desert dust radiative forcing on Sahel precipitation: Relative importance of dust compared to sea surface temperature variations, vegetation changes, and greenhouse gas warming. *Journal of Climate*, **20**, 1445-1467.

4) **Higgins**, R.W., V.B.S. Silva, W. Shi, and J. Larson, 2007: Relationship between climate variability and fluctuations in daily precipitation over the United States. *Journal of Climate*, **20**, 3561-3679.

5) **Xue**, Y., R. Vasic, Z. Janjic, F. Mesinger, and K.E. Mitchell, 2007: Assessment of dynamic downscaling of the continental U.S. regional climate using the Eta/SSiB Regional Climate Model. *Journal of Climate*, **20**, 4172-4193.

6) **De Sales**, F., and Y. Xue, 2006: Investigation of seasonal prediction of the South American regional climate using the nested model system. *Journal of Geophysical Research*, **111**, D20107, doi:10.1029/2005JD006989.

7) **Kim**, Y.J. and G. L. Wang, 2007a: Impact of initial soil moisture anomalies on subsequent precipitation over North America. *Journal of Hydrometeorology*, **8(3)**, 513-533.

8) **Kim**, Y.J. and G. L. Wang, 2007b: Impact of vegetation feedback on the response of precipitation to antecedent soil moisture anomalies over North America. *Journal of Hydrometeorology*, **8(3)**, 534-550.

9) **Liston**, G.E., C.A. Hiemstra, K. Elder, and D.W. Cline, 2008: Meso-cell study area (MSA) snow distributions for the Cold Land Processes Experiment (CLPX). *Journal of Hydrometeorology*, **9(5)**, TBD.

10) **Hong**, S., V. Lakshmi, and E. Small, 2007: Relationship between vegetation biophysical properties and surface temperature using multi-sensor satellite data. *Journal of Climate*, **20**, 5593-5606.

11) **McNulty**, S.G., G. Sun, E. Cohen, J. Moore-Myers, and D. Wear, 2007: Change in the southern U.S. water demand and supply over the next forty years. In: *Wetland and Water Resource Modeling and Assessment: a Watershed Perspective* [Jin, W. (ed.)]. CRC Press, Taylor & Francis Group, 312 pp.

12) **Del Genio**, A.D., M.-S. Yao, and J. Jonas, 2007: Will moist convection be stronger in a warmer climate? *Geophysical Research Letters*, **34**, L16703, doi:10.1029/2007GL030525.

13) **Millar**, C.I., and R.D. Westfall, 2008: Rock glaciers and related periglacial landforms in the Sierra Nevada, CA, USA: inventory, distribution, and climatic relationships. *Quaternary International*, corrected proof, doi:10.1016/j.quaint.2007.06.004.

4 | Land-Use and Land-Cover Change

Strategic Research Questions

6.1 What tools or methods are needed to better characterize historic and current land-use and land-cover attributes and dynamics?

6.2 What are the primary drivers of land-use and land-cover change?

6.3 What will land-use and land-cover patterns and characteristics be 5 to 50 years into the future?

6.4 How do climate variability and change affect land use and land cover, and what are the potential feedbacks of changes in land use and land cover to climate?

6.5 What are the environmental, social, economic, and human health consequences of current and potential land-use and land-cover change over the next 5 to 50 years?

See Chapter 6 of the *Strategic Plan for the U.S. Climate Change Science Program* for detailed discussion of these research questions.

The global climate system is affected by land-use and land-cover changes through biogeophysical, biogeochemical, and energy exchange processes. These changes in turn affect climate at local, regional, and global scales. Key processes include uptake and release of greenhouse gases by the land cover of the terrestrial biosphere to and from the atmosphere through photosynthesis, respiration, and evapotranspiration; the release of aerosols and particulates from surface land-cover change perturbations; variations in the exchange of sensible heat between the surface and atmosphere due to land-cover changes; variations in absorbance and reflectance of radiation as land-cover changes affect surface albedo; and surface roughness effects on atmospheric momentum that are land cover-dependent. While human activity can alter many of these processes, weather, climate, and geological processes are also important.

For example, changes in land cover modify the reflectance of the land surface, determining the fraction of the Sun's energy absorbed by the surface and thus affecting heat and moisture fluxes. These processes also alter vegetation transpiration and surface hydrology and determine the partitioning of surface heat into latent and sensible heat fluxes. At the same time, vegetation and urban structure determine surface roughness, thus air momentum and heat transport. In addition, deforestation and forest fires alter ecosystems and release carbon dioxide (CO_2), methane (CH_4), carbon monoxide, and aerosols to the atmosphere.

Land-use and land-cover change studies also provide valuable information for large-scale vegetation biomass and forest cover assessments that are key components of the carbon cycle. Future land-use and land-cover change goals include (1) very accurate biomass estimates, thus refining knowledge of carbon storage in vegetation, (2) understanding regional land-use changes that affect biomass, and (3) quantifying linkages and feedbacks between land-use and land-cover change, climate change forcings, climate change, and other related human and environmental components.

Research that examines historic, current, and future land-use and land-cover change, its drivers, feedbacks to climate, and its environmental, social, economic, and human health consequences is therefore of great importance for understanding climate and requires interagency and intergovernmental cooperation. One example of a multi-agency effort is the Congo Basin Forest Partnership, which focuses on conserving this rainforest in equatorial Africa, the second largest tropical rainforest in the world. Satellite data are used to map forest extent, determine habitat fragmentation, enforce conservation laws, and thus minimize greenhouse gas emissions from deforestation land-use changes. Another example is the North American Carbon Initiative that seeks to understand the carbon cycle for North America.

HIGHLIGHTS OF RECENT RESEARCH

The following are selected highlights of recent research into land-use and land-cover change issues supported by CCSP-participating agencies.

A Basin-Scale Econometric Model for Projecting Future Amazonian Landscapes.[1,2] A team of U.S. and Brazilian researchers collaborated in a study predicting the intensity of deforestation of the Amazon Basin resulting from four different economic development

scenarios. Projections were generated from results of econometric modeling based on economic theory and detailed local observation. The projections considered scenarios defined by three factors, including trends in population growth (expected and low rates), the nature of anticipated infrastructure investments (Avança Brasil and successor projects, or not), and efforts at governance (unofficial road building), reflected by the depth of control over forest conservation in protected areas and on private holdings. The study considered potential variations among these three factors to predict the relative percentage of closed forest cleared by 2020. For the scenarios chosen, the resulting projected percentage deforestation (in parentheses) are as follows:

- Expected population growth/Avança Brasil investments/no governance (31%)
- Expected population growth/Avança Brasil investments/medium governance (19%)
- Expected population growth/no Avança Brasil investments/medium governance (19%)
- Low population growth/no Avança Brasil investments/complete governance (16%).

Of importance to the debate about road construction in the Amazon are the results with and without Avança Brasil, for expected levels of population growth and medium governance. Specifically, hardly any difference in deforestation in 2020 is observable, which indicates the minimal impact of the infrastructure projects as defined.

North American Vegetation Dynamics Observed with Multi-Resolution Satellite Data.[3]
Normalized difference vegetation index (NDVI) data from Advanced Very High-Resolution Radiometer (AVHRR) instruments were used from 1982 to 2005 to identify regions in North America that experienced increases in annual photosynthetic capacity from 1982 to 2005. The identified regions were next investigated with Landsat imagery, Ikonos data, aerial photography, and ancillary data to determine the cause of the increase in photosynthesis. Not surprisingly, a range of causes for the NDVI increases were found: increased precipitation; warmer spring conditions; severe drought and subsequent recovery; expansion of irrigated agriculture; logging and subsequent regeneration; and forest fires with subsequent regeneration. Higher latitude areas were affected solely by the climatic influences of warmer temperatures. In other areas, however, land-use and land-cover changes were responsible for

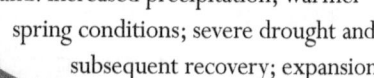

the changes in photosynthesis observed. In the Southern Great Plains, a semi-arid region, a massive expansion of center-pivot irrigated agriculture occurred that was responsible for the regional changes observed (see Figures 10 and 11). Large-scale logging, referred to as "progressive clear cuts," was found to be responsible for significant changes in photosynthesis in the province of Quebec. The logged areas first experienced a decrease in photosynthesis immediately following clearing, followed by a gradual recovery of photosynthesis over the next 10 to 15 years.

This work shows the value of using different types of satellite data to study climate and land-use and land-cover change. Coarse-resolution time series "survey" data are used to identify areas where variations in photosynthesis have occurred, then Landsat

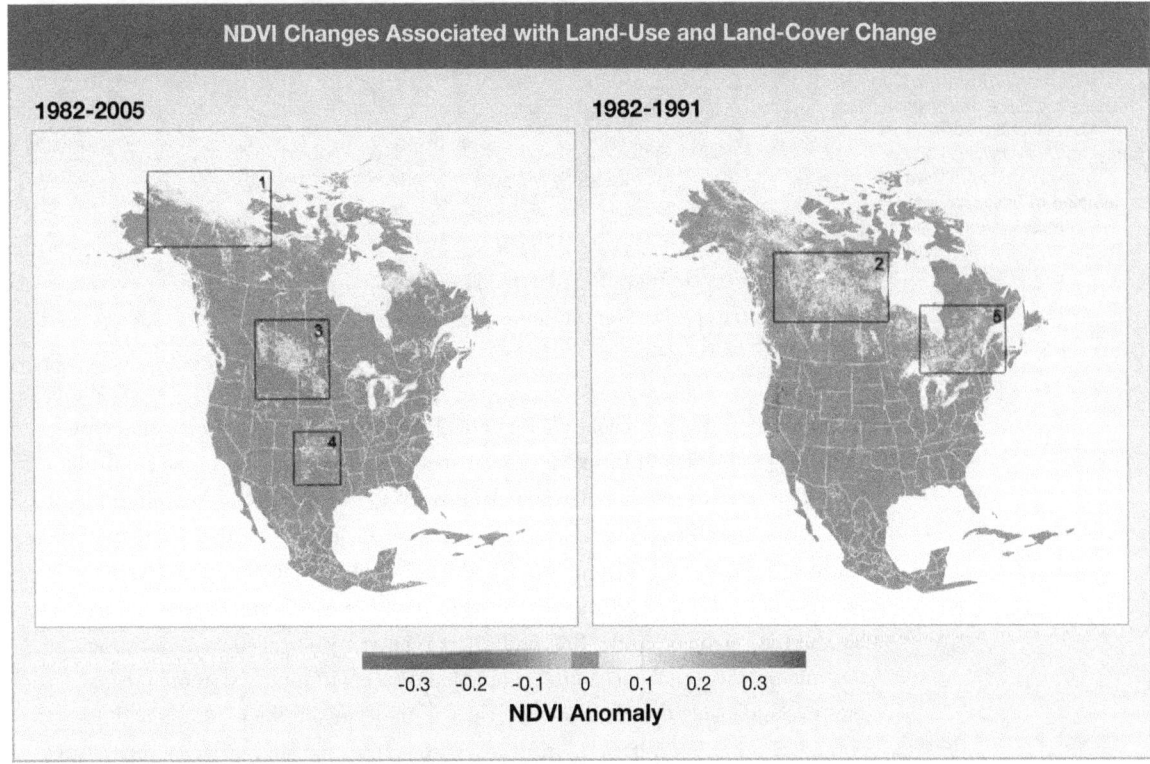

Figure 10: NDVI Changes Associated with Land-Use and Land-Cover Change. Areas of marked changes in NDVI and thus net primary production associated with warmer surface temperatures, increased precipitation, and/or land-use and land-cover changes from 1982 to 2005 (left) and 1982 to 1991 (right) using data with an 8 by 8-km resolution from the noted time periods. In the right figure, the increased NDVI values observed from 1982 to 1991 were due to the following: area 2 experienced warmer temperatures and a longer growing season, coupled with recovery from forest fires and logging in affected areas (the red colors); and area 5 shows the results of large-scale logging. In the left figure, NDVI increases were due to the following: area 1 experienced warmer temperatures and a longer growing season; area 3 experienced warmer temperatures and increased precipitation; and area 4 experienced widespread land-use changes, where dry land farming was replaced by center-pivot irrigation. See Figure 11 for an explanation of the land-use changes that can be observed using coarse-resolution satellite data in area 4. *Credit: C.S.R. Neigh, C.J. Tucker and J.R.G. Townshend, NASA / Goddard Space Flight Center and University of Maryland (reproduced from* **Remote Sensing of Environment** *with permission from Elsevier).*

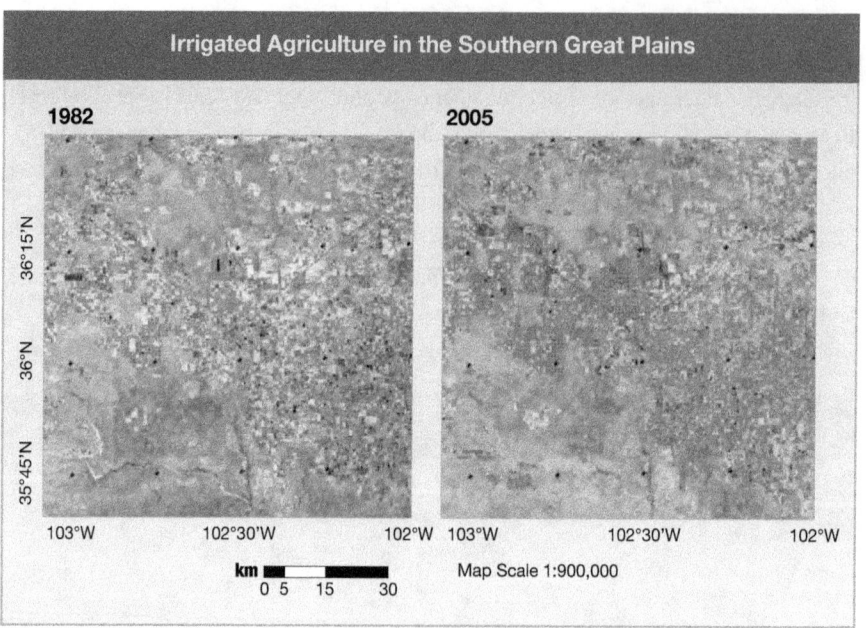

Irrigated Agriculture in the Southern Great Plains

1982

2005

36°15'N

36°N

35°45'N

103°W   102°30'W   102°W   103°W   102°30'W   102°W

km

0 5   15   30

Map Scale 1:900,000

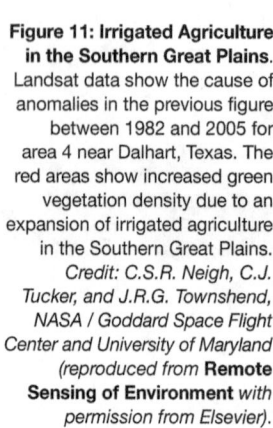

Figure 11: Irrigated Agriculture in the Southern Great Plains. Landsat data show the cause of anomalies in the previous figure between 1982 and 2005 for area 4 near Dalhart, Texas. The red areas show increased green vegetation density due to an expansion of irrigated agriculture in the Southern Great Plains. *Credit: C.S.R. Neigh, C.J. Tucker, and J.R.G. Townshend, NASA / Goddard Space Flight Center and University of Maryland (reproduced from* **Remote Sensing of Environment** *with permission from Elsevier).*

and climate data are used to understand possible land-use changes and/or climatic variation in the specific "survey" areas identified.

Identifying Land-Use and Land-Cover Change in Central America.[4,5,6] Seasonal tropical forests of the southern Yucatan Peninsula form the largest expanse of this ecosystem type remaining in Mexico. It forms an ecocline between a drier region to the north and humid forest to the south in Guatemala. Increasing population and intensification of agriculture since the 1960s has raised international concerns about the effects of these land-use changes on this large carbon and biodiversity reservoir. This led to the creation of the Calakmul Biosphere Reserve to preserve this unique forested area. The Southern Yucatan Peninsular Region Project is currently engaged in an assessment of the vulnerability/resilience of the coupled human-environment system in the face of multiple and highly dynamic land-use changes underway in the region. It addresses the consequences to ecosystem services, forest structure, and land surface temperatures and fire potential in the face of increasing settlement and expansion and intensification of agriculture throughout the region. The project employs Landsat, Moderate Resolution Imaging Spectrometer (MODIS), and AVHRR satellite imagery, ecological information, and socioeconomic studies. Recent biodiversity studies have found that the current landscape matrix maintains the biotic diversity of the reserve but this is threatened by the loss of humid forests to the south and the interruption of biotic flow across the ecocline due to habitat fragmentation. Current land uses also threaten overall biomass productivity due to declining nutrient conservation.

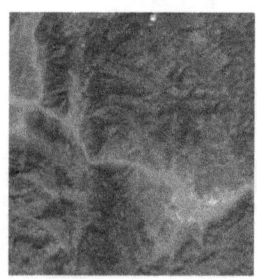

Map of Russian Forest Biomass.[7] Scientists at the Woods Hole Research Center and Oregon State University have collaborated with Russian scientists to produce two maps of forest biomass for Russia. Both maps were based on a regression-tree analytical approach that determined the relationship between ground data collected at 12 sites across Russia and satellite data for all the forests of Russian as part of a forest inventory study in 2000. The total Russian forest biomass estimates ranged from 46 to 67 Gt dry matter. This is important initial work to determine carbon stocks in Russian boreal forests.

North American Native Tallgrass Prairie: High Carbon Storage, Rapid Loss from Cultivation, Slow Increase from Restoration.[8,9] Tallgrass prairie is the most extensive grassland type in North America, ranging from Texas to Minnesota and north into Canada, but over 95% of the original prairie has been converted to agricultural row crops. There is increasing interest in restoring these grassland systems for conservation, biodiversity, carbon sequestration, and other conservation goals. A recent study sampling native prairie, converted prairie in row crops, and restored prairie showed that the largest pools of soil organic carbon in the United States are in prairie grass that survived from earlier times and that, on average, cultivation reduced soil carbon by nearly 30%. However, the carbon content of many restored prairies still did not approach that of native prairie after as long as 60 years, and only at the Texas site did restoration result in significantly higher soil carbon. The effects of land-use change were highly site-specific such that the effects of land use on carbon storage in this region defy any broad generalizations.

Mangrove Forest Losses in Tsunami-Affected Area of Asia.[10] It is estimated that mangrove forests, including associated soils, can sequester approximately 1.5 metric tons of carbon per hectare each year, while disturbance or destruction of these forests can lead to the release of much of this carbon as the greenhouse gases CO_2 or CH_4. Time-series Landsat data from 1975, 1990, 2000, and 2005 were used to identify the present distribution, rate of change, and major causes of change in the tsunami-affected countries of South and Southeast Asia. The analyses for the first time show that (1) the region lost 12% of

its mangrove forests from 1975 to 2005 (their present extent is about 1,670,000 ha); (2) annual deforestation during the same period was highest in Burma (~1%) and lowest in Sri Lanka (0.1%); and (3) net deforestation peaked at 137,000 ha during 1990 to 2000, increasing from 97,000 ha during 1975 to 1990, and declining to 14,000 ha during 2000 to 2005. The major causes of deforestation in the region were agricultural expansion (81%), aquaculture (12%), and urban development (2%), but there are major differences between countries. For example, in Burma, illegal logging and fuel wood collection coupled with degradation due to erosion and sedimentation are major factors, while urban development is a dominant factor in Malaysia and aquaculture is the most important cause in Indonesia. Information generated from this study can be used to identify potential rehabilitation sites and priority conservation areas.

HIGHLIGHTS OF PLANS FOR FY 2009

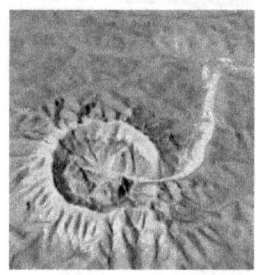

The NASA-USGS Global Land Survey 2005. In order to quantify changes in land cover around the globe, suitable data sets must be made available to the science community. Landsat data are a very suitable source of satellite imagery to identify and map land-use and land-cover changes. A new global satellite data set—the Global Land Survey 2005 (GLS-2005 previously entitled the Mid-Decadal Global Land Survey)—is being prepared by NASA and USGS. The project team is assembling substantially cloud-free Landsat images for the Earth's land areas centered on the 2004 to 2007 period, which are processed into an orthorectified data set. This project follows from previous efforts to assemble global Landsat collections centered on 1975, 1990, and 2000, and will provide a basis for rigorous assessment of land-cover changes during the last several years. Initial GLS-2005 products were released in early 2008, with the complete data set to be available at no cost from USGS by the end of 2008. While relying primarily on Landsat-7 and Landsat-5, GLS-2005 will also include imagery from the Earth Observing-1 (EO-1) Advanced Land Imager (ALI) and Terra Advanced Spaceborne Thermal Emission and Reflection Radiometer (ASTER) instruments as needed. In addition to the data processing activities, a CCSP agency Land-Use and Land-Cover Change initiative is funding a number of peer-reviewed proposals to generate land cover and land-cover change products and to analyze vegetation dynamics from GLS-2005 as well as the older GLS data from the 1970 to 2000 period.

This activity will address CCSP Goals 1 and 2 and Questions 6.1 and 6.2 of the CCSP Strategic Plan.

Land-Use/Land-Cover Change Effects on Soil and Water. USDA research funded through the Cooperative State Research, Education, and Extension Service (CSREES) is developing baseline stratigraphic markers for use in identifying effects of land-use change on riparian soil attributes. The research work will develop a quantitative technique that can be consistently applied over a region to identify layers in riparian soils affected by land-use changes. The ultimate goal is to establish pedologic and floristic changes represented by pollen profiles that are regionally correlated with specific dates and land-use/land-cover changes, and are recognizable from basin to basin. The work is being conducted in the southern New England region of the United States, focusing on Narragansett Bay, the Pawcatuck River, and the Thames River. Other research is developing adaptive agricultural management models that identify the best cropping systems for coping with climate and land-use changes in two watersheds of the Flathead River Basin, Montana. Research objectives include constructing plausible future climate and land-use change scenarios in terms of agriculturally sensitive climate variables, including precipitation, temperature, and CO_2 concentrations. Best cropping systems will be determined in terms of profit or net return per unit land area, and water quality and the rate of soil erosion at the edge of the soil mapping unit. By making the model and associated data sets usable and accessible via an interactive website, agricultural producers will be able to adaptively manage their operations for climate and land-use changes.

This activity will address CCSP Goals 1, 2, 3, and 4
and Questions 6.2, 6.4, and 6.5 of the CCSP Strategic Plan.

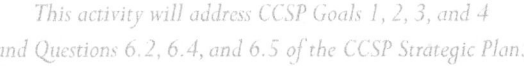

Land-Use and Land-Cover Change Cross-Cuts with Ecosystems and the Global Carbon Cycle.
Newly funded USDA research projects through CSREES cut across various CCSP program elements. A new research project has been initiated to integrate the effects of land-use change on invasive plant species distribution into an invasive plant atlas for the mid-southern United States. The project will quantify the relationships of weed distribution and spread with land use, and use that information directly in educating agriculture stakeholders, natural resource managers, and other interested parties on potential human-induced opportunities for invasive species spread. This project addresses objectives of the Land-Use and Land-Cover Change and Ecosystems research elements.

Another cross-cutting project has been funded to assess the effects of land-cover and land-use change on carbon stocks in the southern United States, giving special attention to translating site-specific carbon pools to landscape scales. This project will investigate soil carbon in relation to land use, land cover, hydrology, topography, and other landscape attributes. The work addresses issues common to the Land-Use and Land-Cover Change and Global Carbon Cycle research elements. It also addresses USDA's priority research areas, including spatially explicit soil carbon modeling.

This activity will address CCSP Goals 1, 2, 3, and 4 and Questions 6.2, 6.3, and 6.4 of the CCSP Strategic Plan.

Impacts and Interactions with Socioeconomic Factors. New projects at NSF will focus on the impacts and interactions among socioeconomic factors, including economic globalization, land-use and land-cover change, and climate change, particularly in Arctic and boreal regions.

This activity will address CCSP Goals 1, 2, and 4 and Questions 6.1, 6.2, and 6.5 of the CCSP Strategic Plan.

LAND-USE AND LAND-COVER CHANGE
CHAPTER REFERENCES

1) **Perz**, S.G., C. Overdevest, M.M. Caldas, R.T. Walker, and E.Y. Arima, 2007: Unofficial road building in the Brazilian Amazon: Dilemmas and models for road governance. *Environmental Conservation,* **34(2)**, 112-121, doi:10.1017/S0376892907003827.

2) **Caldas**, M., R.T. Walker, S. Perz, E. Arima, S. Aldrich, and C. Simmons, 2007: Theorizing land cover and land use change: The peasant economy of colonization in the Amazon Basin. *Annals of the Association of American Geographers*, **97(1)**, 86-110.

3) **Neigh**, C.S.R., C.J. Tucker, and J.R.G. Townshend, 2008: North American vegetation dynamics observed with multi-resolution satellite data. *Remote Sensing of Environment*, **112**, 1749-1772.

4) **Vester**, H.F., D. Lawrence, J.R. Eastman, B.L. Turner II, S. Calme, R. Dickson, C. Pozo, and F. Sangermano, 2007: Land Change in the southern Yucatan and Calamul biosphere reserve: Effects on habitat and biodiversity. *Ecological Applications,* **17**, 989-1003.

5) **Lawrence**, D., P. D'Odorico, L. Diekmann, M. DeLonge, R. Das, and J. Eaton, 2007: Ecological feedbacks following deforestation create the potential for a catastrophic ecosystem shift in tropical dry forest. *Proceedings of the National Academy of Sciences*, **104**, 20696-20701.

6) **Manson**, S.M. and T. Evans, 2007: Land change science special feature: Agent-based modeling of deforestation in southern Yucatán, Mexico, and reforestation in the Midwest United States. *Proceedings of the National Academy of Sciences*, **104**, 20678-20683.

7) **Houghton**, R.A., D. Butman, A. Bunn, O.N. Krankina, P. Schlesinger, and T.A. Stone, 2007: Mapping Russian forest biomass with satellites and forest inventories. *Environmental Research Letters*, **2(4)**, doi:10.1088/1748-9326/2/4/045032.

8) **McCulley**, R.L., N. Fierer, and R.B. Jackson, 2007: Restoration of grasslands after agriculture: Insights from regional chronosequences. *Abstracts of the Ecological Society of America Annual Meeting, 5-10 Aug 2007, San Jose, California.* Available at <eco.confex.com/eco/2007/techprogram/ P1427.HTM>.

9) **McCulley**, R.L., T.W. Boutton, and S.R. Archer, 2007: Soil respiration in a subtropical savanna parkland: Response to water additions. *Soil Science Society of America Journal*, **71**, 820-828.

10) **Giri**, C., Z. Zhu, L.L. Tieszen, A. Singh, S. Gillette, and J.A. Kelmelis, 2008: Mangrove forest distribution and dynamics (1975-2005) of the tsunami-affected region of Asia. *Journal of Biogeography*, **35(3)**, 519–528.

5 | Global Carbon Cycle

Strategic Research Questions

7.1 What are the magnitudes and distributions of North American carbon sources and sinks on seasonal to centennial time scales, and what are the processes controlling their dynamics?

7.2 What are the magnitudes and distributions of ocean carbon sources and sinks on seasonal to centennial time scales, and what are the processes controlling their dynamics?

7.3 What are the effects on carbon sources and sinks of past, present, and future land-use change and resource management practices at local, regional, and global scales?

7.4 How do global terrestrial, oceanic, and atmospheric carbon sources and sinks change on seasonal to centennial time scales, and how can this knowledge be integrated to quantify and explain annual global carbon budgets?

7.5 What will be the future atmospheric concentrations of carbon dioxide, methane, and other carbon-containing greenhouse gases, and how will terrestrial and marine carbon sources and sinks change in the future?

7.6 How will the Earth system, and its different components, respond to various options for managing carbon in the environment, and what scientific information is needed for evaluating these options?

See Chapter 7 of the *Strategic Plan for the U.S. Climate Change Science Program* for detailed discussion of these research questions.

The U.S. Carbon Cycle Science Program is making progress in understanding the changes, magnitudes, and distributions of carbon sources and sinks, the processes operating within and between major terrestrial, oceanic, and atmospheric carbon reservoirs, and the underlying mechanisms involved, including human activities, fossil fuel emissions, land use, and climate forcings. Program scientists are currently quantifying many of the intricate complexities and interactions between the major

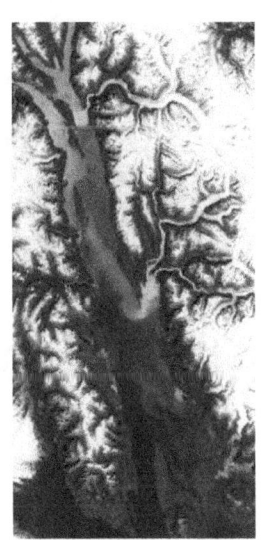

U.S. CARBON CYCLE SCIENCE PROGRAM

The U.S. Carbon Cycle Science Program contributes to all goals of the *CCSP Strategic Plan* (2003)—focusing particularly on Goal 2, "Improved quantification of the forces bringing about changes in the Earth's climate and related systems." The program addresses directly the six overarching carbon cycle questions of Chapter 7 of the *CCSP Strategic Plan*. The research element is synergistic with the Ecosystems, Global Water Cycle, Climate Variability and Change, Atmospheric Composition, Land-Use and Land-Cover Change, and Human Contributions and Responses research elements. The agencies responsible for carbon cycle research are DOE; NASA; NIST; NOAA; NSF; USDA's Agricultural Research Service (ARS), Cooperative State Research, Education, and Extension Service (CSREES), Forest Service, and Natural Resources Conservation Service (NRCS); and USGS. Together, they have planned and are coordinating a multidisciplinary research strategy to integrate the broad range of needed infrastructure and resources, scientific expertise, and stakeholder input essential for program success and improved decision processes.

carbon reservoirs and climate. To execute this undertaking, Federal agencies and departments with carbon cycle interests coordinate, manage, and support the overall science and implementation plans under two major thrusts: North American Carbon Program (NACP) and Ocean Carbon and Climate Change (OCCC) Program. As these science programs mature and generate needed carbon observations, field and experimental results are being used to constrain advanced carbon models at scales from experimental sites to regions as an important means of incorporating site, regional, and global observations into global carbon models and analyses. The ultimate objective is to develop increasingly realistic and predictive coupled carbon-climate and Earth system models to provide better insight into future feedbacks and drivers between the major components of the Earth system.

In FY 2009, a new CCSP-wide research priority will be initiated to quantify the magnitude and dynamics of carbon cycling of high-latitude ecosystems under abrupt climate change. In support of this research initiative, the U.S. Carbon Cycle Science Program will coordinate a concerted Federal effort addressing high-latitude carbon cycle research (observations, attribution, prediction, and

THE NORTH AMERICAN CARBON PROGRAM

NACP is designed to address strategic research question 7.1, and elements of questions 7.2 through 7.6, in Chapter 7 of the *CCSP Strategic Plan*. For example, it will quantify the magnitudes and distributions of terrestrial, freshwater, oceanic, and atmospheric carbon sources and sinks for North America and adjacent coastal oceans; enhance understanding of the processes controlling source and sink dynamics; and produce consistent analyses of North America's carbon budget that explain regional and continental contributions and year-to-year variability. This program is committed to reducing uncertainties related to the increase of carbon dioxide and methane in the atmosphere and the amount of carbon, including the fraction of fossil fuel carbon, being taken up by North America's ecosystems and adjacent coastal oceans.

THE OCEAN CARBON AND CLIMATE CHANGE PROGRAM

OCCC is designed to address strategic research question 7.2, and elements of questions 7.3 through 7.6, in Chapter 7 of the *CCSP Strategic Plan*. For example, in regards to question 7.2, it will focus on oceanic research aimed at quantifying how much atmospheric carbon dioxide is being taken up by the ocean at the present time and how climate change will affect the future behavior of the oceanic carbon sink. The terrestrial and ocean carbon programs are synergistic, integrating program activities addressing carbon dynamics on the coastal shelves adjacent to North America (questions 7.1 and 7.2), where carbon changes in the terrestrial system greatly influence carbon processes in the coastal ocean.

mitigation), which will be conducted in unison with its priorities under the NACP and OCCC programs. To accomplish the carbon cycle element of the new CCSP priority at high latitudes, the interagency working group will solicit new investments and reprogram previous research investments to complement current research in order to fill gaps, and promote and augment ongoing carbon observations and networks in high-latitude lands and ocean ecosystems. The enhanced emphasis on high-latitude ecosystems will provide critical scientific information on past and current carbon dynamics of undersampled regions of North America and adjacent oceans, as well as other undersampled regions of the world, such as Antarctica and the adjacent Southern Ocean.

As research programs mature, scientific and governmental collaborations on carbon cycle science are broadening and escalating with international neighbors within North America as well as with extended Northern Hemisphere interests, international organizations, and global partners.

HIGHLIGHTS OF RECENT RESEARCH

The research highlights that follow are recent selected accomplishments of the U.S. Carbon Cycle Science Program. These accomplishments span carbon cycle issues related to climate forcing factors, terrestrial and oceanic sinks and sources, the atmospheric reservoir, global carbon analysis, carbon management, and other relevant biogeochemical exchanges between the major Earth reservoirs that link to climate.

Climate Forcing

Carbon dioxide (CO_2) and methane (CH_4) are significant forcing agents of climate, and their atmospheric concentrations have been increasing over the past 2 centuries, attributed primarily to human activities. Approximately 85 to 90% of present-day anthropogenic emissions are attributed to fossil fuel combustion, with land-use change accounting for most of the rest. Future concentrations of CO_2 and CH_4 in the atmosphere will depend on the long-term trends in terrestrial and oceanic reservoirs, the rate of exchange between Earth reservoirs, the variability in natural and anthropogenic emissions, and the capacity of natural and managed sinks.

Factors Affecting Fossil Fuel Emissions.[1] An analysis was completed of factors that influence the magnitude, regional patterns, and temporal trends of global emissions of CO_2 from fossil fuels, which dominate climate change forcing. The analysis included demographic, economic, and technological drivers of fossil fuel emissions by using annual time-series data on national emissions, population, energy consumption, and gross domestic product (GDP). Fossil fuel CO_2 emissions can be represented as the product of four driving factors: global population, per capita world GDP, energy intensity of world GDP, and carbon intensity of energy. Results show that growth of global CO_2 emissions since 2000 was driven by a cessation or reversal of earlier declining trends in the energy intensity of GDP and the carbon intensity of energy, coupled with continuing increases in population and per capita GDP. Nearly constant or slightly increasing trends in the carbon intensity of energy have recently been observed in both developed and developing regions.

Terrestrial Carbon Cycle

The terrestrial carbon and water cycles comprise a complex set of interactive biogeochemical processes that transfer carbon between land, oceans, and the atmosphere. Collectively, these processes influence atmospheric CO_2 and CH_4 concentrations. Improving the scientific understanding of the role of these reservoirs and processes in the carbon cycle reduces uncertainty about the factors influencing greenhouse gas increases and provides a stronger foundation for climate change decision support, in particular for carbon management to mitigate CO_2 and CH_4 increases in the atmosphere.

Northern Hemisphere Terrestrial Carbon Sink Analysis.[2,3] Temperate and boreal forests in the Northern Hemisphere act as a substantial carbon sink of 0.6 to 0.7 GtC yr^{-1}, yet recent results from the AmeriFlux research network show that forest disturbance from harvest and fire are responsible for much of the overall variability in forest carbon sequestration. Forests are a carbon source to the atmosphere for as many as 20 years after these events, followed by a long period of carbon sequestration. Using results from observation networks in the United States and Europe, a recently completed Northern Hemisphere synthesis of net ecosystem exchange of CO_2 from differently aged forests found that

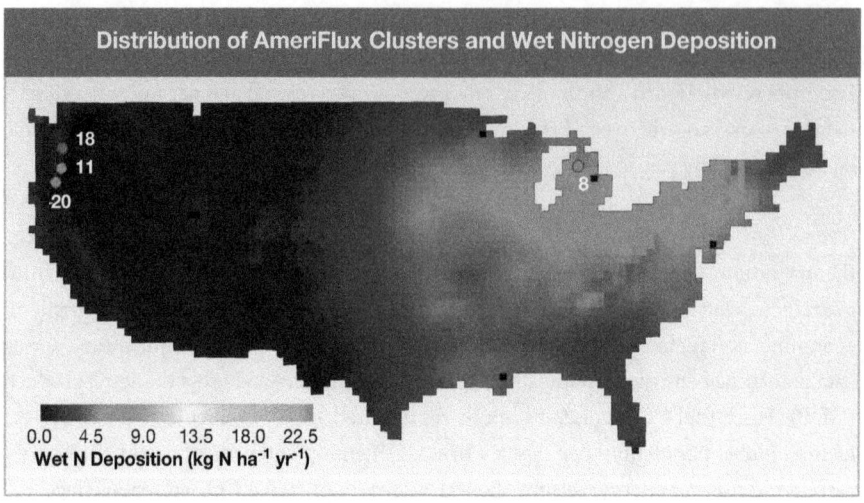

Distribution of AmeriFlux Clusters and Wet Nitrogen Deposition

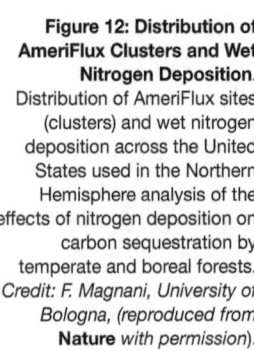

Figure 12: Distribution of AmeriFlux Clusters and Wet Nitrogen Deposition. Distribution of AmeriFlux sites (clusters) and wet nitrogen deposition across the United States used in the Northern Hemisphere analysis of the effects of nitrogen deposition on carbon sequestration by temperate and boreal forests. *Credit: F. Magnani, University of Bologna, (reproduced from* **Nature** *with permission).*

forest age, as a function of stand-replacing disturbances, accounted for approximately 90% of the total variability in net carbon sequestration. The average net carbon uptake over the harvest cycle of the forests was 56% of the maximum observed in mature forests. After accounting for age and disturbance effects (wildfires, harvesting, infestations, etc.), low continuous levels of nitrogen deposition (up to 10 kg N ha^{-1} yr^{-1} wet deposition), largely the result of anthropogenic activities, appear to overwhelmingly account for additional carbon sequestration by these forests (see Figure 12).

Carbon Distribution between Forest Root and Shoot Systems.[4] One of the largest uncertainties in estimating changes in carbon stocks and understanding the effects of global change on forest carbon sequestration involves carbon allocation to coarse tree roots. Using an approach incorporating successional dynamics in plant communities and whole-tree

harvesting including root excavation, a dynamic pattern of root-to-shoot ratios was revealed. Ratios varied from 0.17 for sites with trees less than 5 years old, to 0.80 for a site with 8-year old trees, to 0.29 for a site with 55-year old trees. Determining the causes of this variability in the root-to-shoot biomass allocation through forest maturation requires further research on how these patterns change as functions of growth, environment, and management. Carbon allocation in forest systems has important implications for projecting belowground net primary production responses to global change in studies of regional and continental carbon fluxes.

Carbon and Nitrogen Cycles of Terrestrial Ecosystems.[5] **One** of the most central processes in the global carbon cycle is the breakdown of plant litter—dead leaves, branches, roots—that releases CO_2 to the atmosphere and provides nitrogen in various forms that can support new plant growth. Understanding what controls the rate of breakdown and nitrogen release in different environments is critical for predicting the effects of climate change on the carbon balance of terrestrial ecosystems. The results of a 10-year experiment conducted at 21 sites showed that in all ecosystems except dry grasslands the amount of nitrogen released is controlled by the initial concentration of nitrogen in the litter and the mass remaining. The results were used to produce simple equations to predict nitrogen release. Because nitrogen is intimately involved in controlling both plant growth and litter breakdown, these equations will contribute to better models of the global carbon cycle.

Oceanic Carbon Cycle

The global ocean is a large and important carbon reservoir that regulates the uptake, storage, and release of CO_2, CH_4, and other climate-relevant chemical species to the atmosphere. The future biogeochemical behavior of this reservoir is uncertain because of potential anthropogenic impacts on many ocean processes, in particular the impact of ocean circulation on carbon exchange and the impact of ocean acidification on the physiology, function, and structure of the complex and diverse ocean ecosystem.

Atmospheric Impact on the Ocean Carbon Reservoir.[6,7] The absorption of anthropogenic CO_2 and the deposition of acid rain from fossil fuel and agriculture emissions can both contribute to the acidification of the global ocean, altering surface seawater acidity, and inorganic carbon storage. Researchers have compared these inputs and concluded that (1) acid rain contributes a minor amount (2%) of acidity compared to the ocean uptake of anthropogenic CO_2, although this value likely represents an upper limit, and (2) the decrease in surface alkalinity from acid rain drives a net air-sea release of CO_2, reducing surface dissolved inorganic carbon (DIC). Total alkalinity and DIC changes mostly offset each other, resulting in a small increase in surface acidity. Additional impacts arise from atmospheric nitrogen deposition, leading to elevated primary production and biological drawdown of DIC that in some places reverses the sign of the surface acidity and air-sea CO_2 exchange. On a global scale, the alterations in surface water chemistry from anthropogenic deposition are a few percent of the acidification, although the impacts are more substantial in coastal waters, where the ecosystem responses to ocean acidification could have the most severe implications for humans.

High-Latitude Systems

High-latitude systems are becoming increasingly important sources of CO_2 and CH_4 to the atmosphere as regional warming changes ecosystem dynamics in the cold regions. Understanding carbon dynamics in high-latitude systems and the factors that may lead to changes in those dynamics are crucial elements of global carbon modeling and essential for understanding the linkages and feedbacks between carbon reservoirs, ecosystems, land cover, hydrology, and climate variability.

Significance of Marginal Ice Zones.[8] Within the Southern Ocean lie regions where biological dynamics are low, called High Nutrient, Low Chlorophyll (HNLC) regimes. When adjacent to seasonal sea-ice retreat, these marginal ice zones produce relatively high chlorophyll concentrations, indicative of phytoplankton production, extending thousands of kilometers to sea. These high chlorophyll anomalies are extremely variable temporally and spatially because the size and location of the marginal ice zone, defined as areas of recent ice melting and retreat, are highly variable between seasons and years. The production of phytoplankton biomass within this zone is an important food source for higher trophic levels and significantly affects carbon cycling in and across the region. The magnitude and distribution of these regimes with greatly elevated production were unknown prior to the satellite era. Since then the elevated chlorophyll and production associated with these regions have been documented with Sea-Viewing Wide Field-of-View Sensor (SeaWiFS) satellite data for the Southern Ocean, particularly within the marginal ice zone of the Southern Ocean where melting ice stabilizes the water column leading to shallower mixed layers and the release of critical elements, such as iron, to the water column. Both processes lead to production of phytoplankton, which have an unambiguous impact on carbon cycling and eventual export to the deep sea, where it remains sequestered for a very long time.

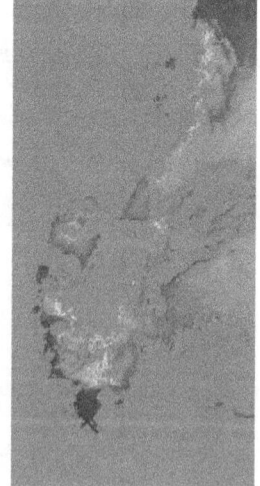

Permafrost Thaw Releases Additional Carbon and Water to Arctic Streams.[9] Pursuing the hypothesis that permafrost thaw and increased infiltration could potentially increase terrestrial respiration of dissolved organic carbon (DOC) and decrease DOC export, researchers investigated historical stream flow records from the Yukon River Basin in Alaska and Yukon Territory with the goal of isolating and quantifying permafrost thaw and/or glacial and perennial snowpack melt effects on the basin water cycle. The analysis quantified a basin-wide increase in groundwater contribution to streamflow of 0.7 to 0.9% per year, but did not find any compelling evidence for a change in total annual water discharge by the basin's rivers. The Yukon River annually discharges approximately 50 km³ of groundwater-derived flow to the Bering Sea. The increased groundwater contribution is consistent with the increased infiltration and DOC

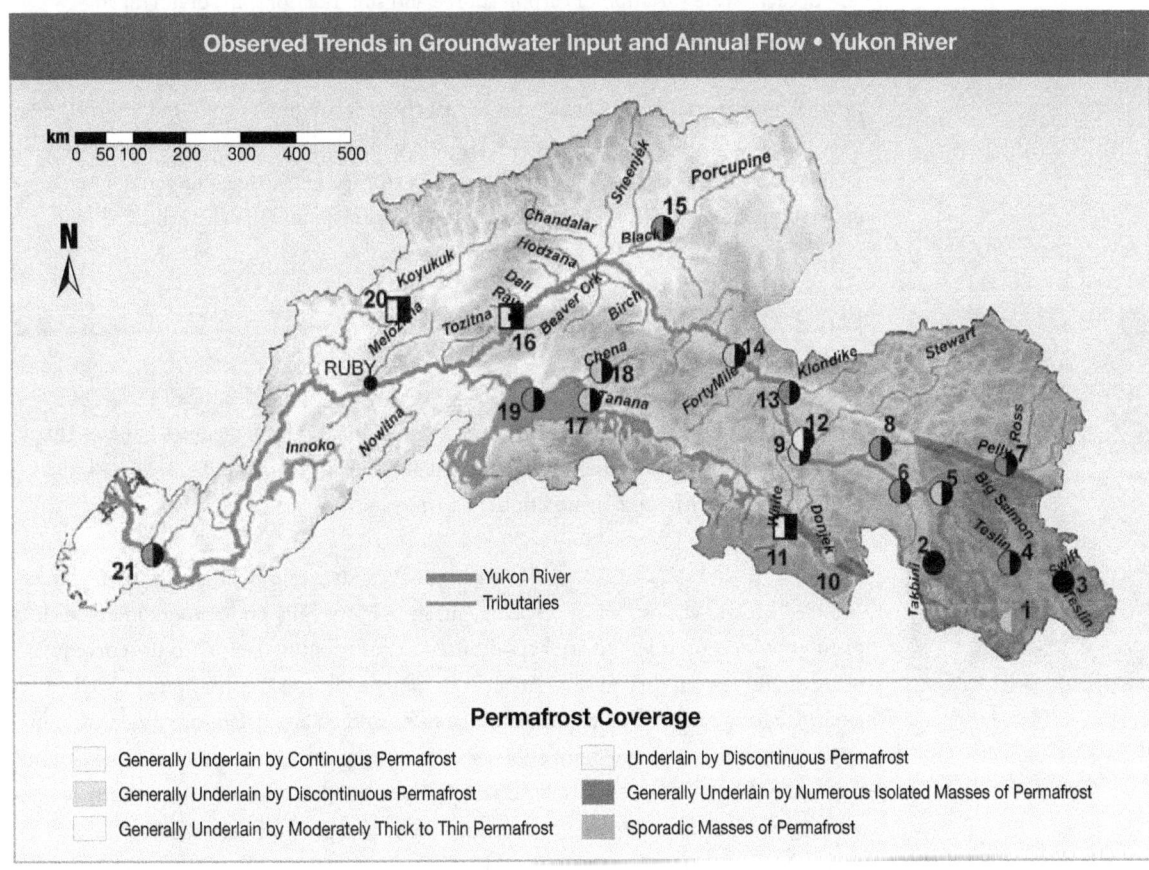

Figure 13: Observed Trends in Groundwater Input and Annual Flow • Yukon River. Observed trends in groundwater input (denoted by left side of symbol) and annual flow (denoted by right side of symbol) at Yukon River Basin streamflow stations. Circles and squares indicate flow records longer and shorter than 35 years, respectively. Symbol color scheme indicates statistical significance of Mann-Kendall trend analysis: red = very highly significant (P<0.01) increasing trend, orange = highly significant (0.01<P<0.05) increasing trend, yellow = moderately significant (0.05<P<0.1) increasing trend, light blue = moderately significant (0.05<P<0.1) decreasing trend. *Credit: M.A. Walvoord and R.G. Striegl, USGS (reproduced from* **Geophysical Research Letters** *with permission from the American Geophysical Union).*

consumption hypothesis and supports load calculations that indicate a downward trend in the relationship between water and carbon export by the Yukon River during summer and autumn (see Figure 13).

Wildfire Disturbances in High-Latitude Terrestrial Ecosystems.[10,11,12] Wildfire is a common occurrence in northern high-latitude ecosystems, and the ecosystem changes have consequences for carbon feedback to the climate system. Researchers, using a process-based terrestrial ecosystem model, assessed the influence of increases in atmospheric CO_2, climate variability and change, and change in fire disturbance on the exchange of CO_2 and CH_4 in high-latitude terrestrial ecosystems. Using historical fire records through 2002, model analysis indicates that fire played a central role in interannual- and decadal-scale variation of carbon source and sink relationships of northern ecosystems and also suggests that increases in atmospheric CO_2 may be important to consider in addition to changes in climate and fire disturbance. Model projections for northern terrestrial ecosystems indicate that these ecosystems could lose up to 50 GtC over the next 100 years and that net CH_4 emissions could double by the end of the 21st century. These studies suggest that carbon storage in northern terrestrial ecosystems is vulnerable to projected changes in climate and fire disturbance.

Global Carbon Analysis

Ocean phytoplankton and land plants are presently absorbing about half the carbon emissions that humans produce. However, recent global carbon analyses indicate that Earth's reservoirs are becoming less efficient at absorbing fossil fuel emissions and losing their ability to take up additional CO_2.

Carbon Cycle Feedbacks to Climate.[11] How carbon cycle dynamics will change with climate and feed back to atmospheric CO_2 concentration is not fully understood. Increases in plant growing season length are hypothesized as contributing factors to the current observed terrestrial carbon sink. An analysis of growing season variation of CO_2 exchange for a range of vegetation sites, from evergreen and deciduous forests to crop to grasslands and including both cool-season and warm-season vegetation types, found that while the growing season length affected how much CO_2 could be potentially assimilated by a plant community over the course of a growing season, other factors such as nutrient and water availability were also important at this scale. This implies

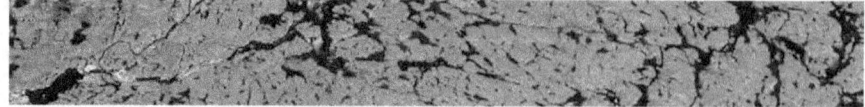

that the climate warming-induced increase in growing season length may have a limited enhancement effect on terrestrial carbon uptake.

Amazon Forests May be More Resilient than Predicted.[14] Coupled carbon-climate models predict substantial carbon loss from tropical ecosystems and drought-induced collapse of the Amazon forest. These models include a physiological feedback mechanism whereby transpiration is reduced in response to initial drought, which in turn exacerbates the drought by interrupting the supply of transpired water that would otherwise contribute to "recycled" precipitation that supports forest growth. Satellite observations of intact Amazon forests subjected to a widespread and severe drought in 2005 show that the response of the forest is actually to green up, which is indicative of increased transpiration and carbon uptake. Apparently, these deep-rooted forests are responding to the increased availability of sunlight and not to water limitations. These observations suggest that intact Amazon forests may be more resilient than some current models assume, at least to short-term climatic anomalies. Future studies are needed to address forest responses to longer term drought in order to better understand the conditions under which water limitations may actually trigger reductions in carbon uptake (see Figure 14).

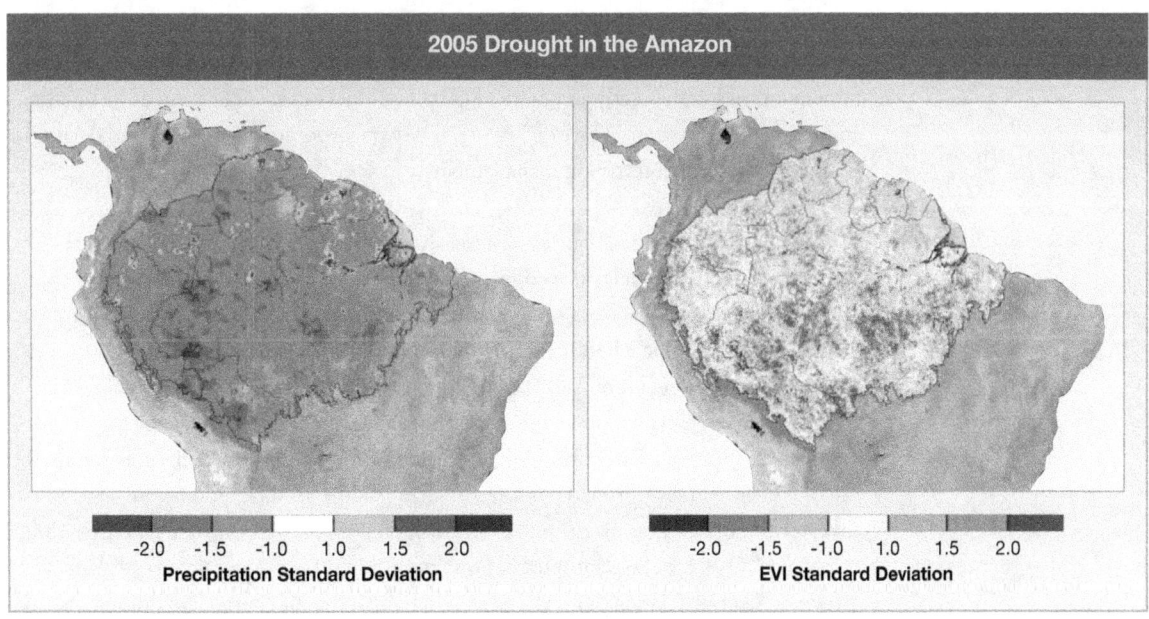

Figure 14: 2005 Drought in the Amazon. During the 2005 drought in the Amazon, intact primary forest showed an increase in photosynthetic activity (right image) despite below-average rainfall (left image). Data from NASA's Terra satellite (right) showed areas of higher (green) and lower (red) growth during the peak of the drought (July-September). Data from the Tropical Rainfall Measuring Mission satellite (left) showed areas of severe rainfall reduction due to the drought (yellow to red) and few areas with above normal rainfall (green to blue). *Credit: K. Didan, University of Arizona.*

Carbon Management and Decision Support

Carbon cycle research provides scientific information to policy and resource decisions about carbon management and mitigation of climate change. The research supported by the carbon cycle program is informing agricultural and resource managers on sequestration, alternative fuels, and inventories, as described in the highlights below. The impact of this research on management strategies is expected to increase over the course of this program.

Soil Carbon Sequestration in Agricultural Lands.[15,16] **Based on** *in situ* soil carbon concentrations, crop growth characteristics, tillage practices, land-use classification using satellite imagery, and climate variables, a geographic information system (GIS)-based Environmental Policy Integrated Climate (EPIC) model projected the amounts of soil carbon sequestered for an agricultural region in sub-Saharan Africa (Mali). Under contemporary ridge-tillage management practices, year-to-year crop variations were attributed primarily to rainfall, the amount of plant-available water, and the amount of fertilizer applied. Under conventional cultivation, with minimal fertilization and no crop residue management, topsoil was continuously lost due to erosion. The model projections suggest that soil erosion is controlled and soil carbon sequestration is enhanced with a ridge-tillage system because of increased water infiltration. The combination of modeling with land-use classification was used to calculate that about 54 kg C ha^{-1} yr^{-1} (5.4 g C m^2 yr^{-1}) may be sequestered in the study area with ridge tillage, increased application of fertilizers, and residue management. The EPIC model is now incorporated in a web-based decision-support system for soil carbon management.

Biofuels from Prairie Grasses.[17] The search for alternatives to fossil fuels has attracted attention to biofuels (fuels derived from plants) and especially to corn as a source of ethanol. A study in Minnesota showed that mixtures of native prairie plants, grown on degraded land, may be a better source of biofuel than corn ethanol or soybean biodiesel. Prairie vegetation yields 51% more energy per hectare than is obtained in ethanol from corn grown on fertile land. The higher net energy gain is due in part to much lower inputs such as cultivation, herbicides, irrigation, and fertilizer, as well as the use of the entire above-ground plant rather than just the seed. The prairie vegetation also stores more carbon over time in soils than do systems growing annual crops, thus removing CO_2 from the atmosphere. Low-input, high-diversity prairie vegetation grown on degraded lands can serve as an efficient source of energy and does not compete for fertile soils needed for food production.

An Assessment of the North American Carbon Budget.[18] An evaluation of North American carbon sources and sinks was generated as part of CCSP's Synthesis and Assessment

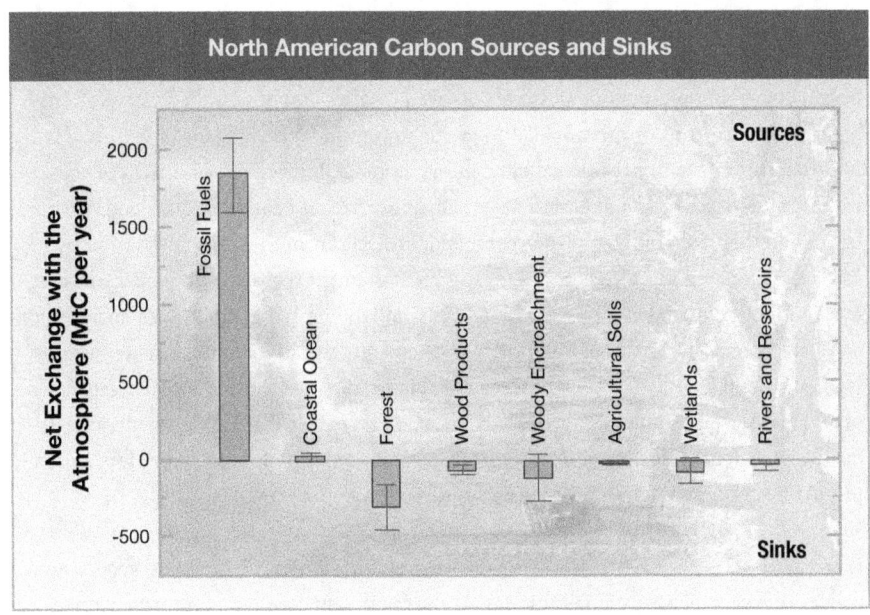

Figure 15: North American Carbon Sources and Sinks. North America is currently a net carbon source (1,336 ± 334 MtC yr^{-1}). A net terrestrial sink of 520 ± 260 MtC yr^{-1} is equivalent to about 30% of fossil fuel emissions in 2003. *Credit: CCSP Synthesis and Assessment Product 2.2.*

Product 2.2, *The First State of the Carbon Cycle Report (SOCCR): North American Carbon Budget and Implications for the Global Carbon Cycle.* The report quantifies North America's fossil fuel emissions for 2003 as 1,856 million metric tons of carbon ±10% (27% of global emissions) and its land sink, primarily in U.S. forests, as 500 million metric tons of carbon ±50% (approximately 30% of North America's emissions). The net release to the atmosphere is 1,350 million metric tons of carbon per year ±25% (see Figure 15). A key finding of the report is that this difference between the sources and sinks is expected to become larger, and that actions to address the imbalance will likely require options that include both emissions reduction and sink enhancement.

North American Carbon Exchange.[19] A CO_2 reanalysis system, CarbonTracker, was used to determine global surface sources and sinks of CO_2. The reanalysis extends from 2000 to 2006 (see <carbontracker.noaa.gov>). Weekly estimates of surface fluxes were produced for 221 land and ocean regions. This advanced data assimilation scheme is focused on relatively well-observed regions, and uses a two-way, nested transport model over North America to simulate circulation at relatively high resolution. The system is capable of handling large amounts of data, and in the most recent analysis over 28,000 individual atmospheric CO_2 observations from the U.S. and Canadian Meteorological Services were assimilated. North American land regions were determined to be a net sink of CO_2, with a mean annual uptake of 0.6 ± 0.4 GtC yr^{-1} and interannual variations of about 0.3 GtC yr^{-1}. This offsets almost one-third of the approximately 1.8 GtC yr^{-1} emissions from fossil fuel burning across North America during the same period.

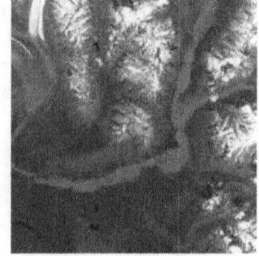

HIGHLIGHTS OF PLANS FOR FY 2009

Enhancing and implementing new carbon cycle studies and observation networks in high-latitude regions of the world along with continuing integration of the NACP and OCCC programs will provide valuable information and improved estimates of the carbon sources and sinks of North American and adjacent coastal systems and ocean basins and their role in the global carbon budget. Data from these observation networks will be assimilated in more comprehensive and advanced regional and global carbon cycle models, coupled carbon-climate models, and integrated Earth system analyses for assessing potential impacts of fossil fuel emissions on terrestrial and ocean ecosystems, land cover and land use, and carbon management strategies. With improved estimates of and greater certainties about the major carbon reservoirs on Earth, scientists will have new insight on how Earth systems functioned under past and present forcings, and predict better how they will respond to future climate forcings.

High-Latitude Carbon Cycle Research. Peatlands (regions of partially decayed vegetation matter) cover a relatively small fraction of the Earth's surface but store nearly one-third of global soil carbon. Climate warming in interior Alaska is already causing some northern peatlands to dry out, while other northern areas are becoming wetter. These changes could have a major effect on the carbon balance of peatlands, including the potential for the release of CO_2 and CH_4, which are important greenhouse gases.

- *Impact of Global Warming* – To determine how climate warming is likely to affect the release of CO_2 and CH_4 from Arctic peatlands, a multi-discipline and -site temperature and water experiment will be implemented in Alaska's high-latitude ecosystems to change soil temperatures, and both increase and decrease the water table depth.

- *Asik Long-Term Study Site* – Ongoing since 1990, research at the Asik watershed, Noatak National Preserve, northwest Alaska, will continue, focusing on quantifying linkages between the topographic (e.g., taiga and tundra areas) declines in snowpack moisture and duration, changes in soil temperature and moisture, release of soil trace gases (CO_2, CH_4, nitrous oxide), and increased export to the aquatic ecosystem of dissolved inorganic nitrogen, DOC, and dissolved organic nitrogen. In FY 2009, a research emphasis will be on linking observations, process studies, and modeling studies from individual watersheds to regional and synoptic scales, and to continue downscaling models to scales appropriate for land management concerns.

- *Improving High-Latitude Carbon Modeling* – Existing models will continue to be used to synthesize and integrate understanding of 20th-century carbon dynamics for high-latitude regions (north of 45°N) across terrestrial, freshwater, and oceanic ecosystems. In FY 2009, these modeling exercises will be advanced with observational and experimental data to improve further the ability to (1) consider the response of carbon currently frozen in peatlands or permafrost soils, and (2) predict how fire activity and severity will affect carbon storage in northern ecosystems, and in particular the Yukon River Basin.

These activities will address Goals 1, 2, 3, 4, and 5 and Questions 7.1, 7.3, 7.4, and 7.5 of the CCSP Strategic Plan.

North American Carbon Program (NACP). **New and reprogrammed investments will be made in observation networks, data management and synthesis, and integrated modeling studies that contribute to NACP goals and objectivities.**

- *Terrestrial Carbon Modeling* – Changes in the carbon cycle can lead to changes in the atmospheric CO_2 concentration and thus feed back and force changes in climate and the terrestrial ecosystem carbon cycle. Describing, attributing causes, and predicting future responses requires a process model; an Integrated Terrestrial Carbon Model (ITCM) will be optimally constrained in structure and function by historical and contemporary observations from AmeriFlux and Free Air CO_2 Enrichment (FACE) data to meet this challenge. Regional-scale ITCM simulations will refine CCSP Synthesis and Assessment Product 2.2 sink estimates by resolving geographic fluxes at a finer temporal resolution. Data assimilation for the NACP Mid-Continent Intensive campaign is underway, and regional-scale simulations for the recent historical period will be analyzed to determine causes of seasonal and interannual variations in carbon source and sink strength.

This activity will address Goals 1, 2, 3, 4, and 5 and Questions 7.1, 7.3, 7.4, 7.5, and 7.6 of the CCSP Strategic Plan.

Ocean Carbon and Climate Change (OCCC). **New and reprogrammed investments will be made in ocean and coastal carbon observations, synthesis, and modeling studies to support OCCC programs, analyze ocean satellite data, and assess ocean and coastal carbon sinks and sources.**

- *Ocean Carbon Observations* – During FY 2009, NSF's Carbon and Water in the Earth System projects—covering regimes spanning the globe, including ocean, coastal, lake, forest, and tundra ecosystems—will expire, while a few ocean and coastal programs will continue beyond FY 2009. New ocean carbon programs will be announced, with submission target dates in FY 2008 and FY 2009, through NSF's Division of Ocean Sciences, NASA's Research Opportunities in Space and Earth Science, and NOAA's Global Carbon Cycling Program. These programs will be supported with reprogrammed investments into a wide range of ocean topics covering carbon cycling, including ocean acidification, terrestrial and coastal carbon exchange, ocean carbon uptake and storage, and related biogeochemical cycles.

- *Large-Scale Coastal Surveys* – Studies are currently underway in the coastal regions of North America to study the incursion and transport of anthropogenic CO_2 and other tracers in coastal ocean waters. This is the first comprehensive effort to measure

partial pressures of CO_2 and related chemical and hydrographic measurements, including DIC, alkalinity, oxygen, nutrients, and CTD (conductivity, temperature, depth), over the entire coastal zone. The survey will establish a baseline describing the geographic distribution of carbon system parameters and develop a set of hydrographic transects with full water column measurements to be revisited over time for studies of interannual changes in physical, chemical, and biological characteristics of the coastal ocean.

- *Ocean Carbon Modeling* – Two key processes are being targeted for improvements in ocean carbon modeling. The first is the role of the Southern Ocean in climate forcing. The research aims to quantify its rates of water mass transfer associated with different circulations, to understand why water mass transformations differ greatly between models, and to analyze the relative contributions to numerical simulation uncertainties from physical and biogeochemical model components. The second is the capacity of the ocean to sequester carbon. The research addresses the nutrient limitation of CO_2 fertilization in the tropical oceans, which are among the regions with the greatest sources of uncertainty in the carbon cycle over the next half century.

These activities will address Goals 1, 2, 3, 4, and 5 and Questions 7.1, 7.2, 7.3, and 7.4 of the CCSP Strategic Plan.

Global Carbon Analysis. The carbon cycle is an integrated global system, and a complete understanding of changes in atmospheric CO_2 and CH_4 concentrations requires a global analysis of carbon sources and sinks and their dynamics at relevant spatial and temporal scales.

- *Global Carbon Cycle Modeling and Analysis* – New research investments for FY 2009 will focus on developing new or improved carbon models that are more comprehensive in treating significant processes and drivers, including those involving or stemming from human activities. These advanced carbon models will address time scales of decades to centuries and integrate across spatial scales up to the global scale. One important focus for improving and extending the treatment of fundamental processes in carbon cycle models is to advance the coupling of global carbon and climate models allowing analyses of interactions and feedbacks within the coupled carbon and climate systems. New research will seek also to extend carbon data assimilation or data fusion schemes by incorporating models of major carbon cycle components with substantially improved detail, realism, and accuracy in the representation of key processes that determine magnitudes and distributions of sources and sinks for carbon and carbon cycle dynamics affecting CO_2 or CH_4 concentrations. New research will be conducted to prepare atmospheric transport and inversion models and data assimilation and data fusion schemes to utilize measurements of atmospheric column CO_2 from the Orbiting Carbon Observatory.

These activities will address Goals 1, 2, 3, 4, and 5 and Questions 7.1, 7.2, 7.3, 7.4, and 7.5 of the CCSP Strategic Plan.

Carbon Management and Decision Support. Initial projects are nearing completion and others are underway that will allow government agencies, industry associations, and private landowners to use carbon management information derived from scientific observations, ecosystem models, and on-line tools (products of carbon cycle science research) for resource management and policy decisions affecting both the near and long term. These projects are have particular importance for forest management and agricultural practices.

- *Agriculture Systems and Emissions of Greenhouse Gases* – A decision-support system for controlling greenhouse gas emissions from agricultural cropping systems will be completed and tested using greenhouse gas emission data from the nationwide Greenhouse Gas Reduction through Agricultural Carbon Enhancement network (GRACEnet). A database of greenhouse gas emissions from crop, pasture, and rangeland systems in the United States will be made publicly available. Remote-sensing technologies for mapping crop residue cover and crop management practices that affect soil conservation and soil carbon will be refined and tested.

- *Monitoring Soil Resources in Agricultural Lands* – Increasing pressure on soil resources in combination with the need to understand how these resources are responding to changing climate has led to the development of a national soil-monitoring network. The monitoring network is currently evaluating soil carbon stock trends in the upper Midwest using a combination of soil sampling, satellite remote-sensing data, and modeling. The results will inform policymakers about carbon sources and sinks in agricultural regions, as well as form a basis for projections of greenhouse gas mitigation potential through carbon management practices. The monitoring network is being established with the USDA Natural Resources Conservation Service National Resources Inventory, which has provided long-term monitoring of land-use and management activity since the early 1980s. The network will provide invaluable information for evaluating the influence of climate forcing and management activity on soil resources, and will ensure long-term sustainability of agricultural production for society.

These activities will address Goals 1, 2, 3, 4, and 5
and Questions 7.1, 7.4, 7.5, and 7.6 of the CCSP Strategic Plan.

International Partnerships. Partnerships between scientists and governments to work on the coordination, synthesis, and interpretation of carbon data sets from around the world are essential for full global-scale syntheses, integrations, and future analysis.

- *North American Partners* – Under the framework of the U.S. Climate Change Bilaterals made from 2001 to 2003 with multiple industrialized and developing countries, government officials from Canada, Mexico, and the U.S. Carbon Cycle Interagency Working Group (CCIWG) recently agreed on a *Statement of Common Interests and Intent to Work Together on Carbon Cycle Research in North America* and endorsed a joint cooperative research effort to assess the North American carbon budget. In 2007, the three countries established the Joint North American Carbon Program and

formed a Joint NACP Science Steering Group and Joint Government Coordination Team to lead and facilitate the common activities of the joint cooperative.

- *Northern Hemisphere Partners* — To extend North American partnership efforts of the CCIWG beyond North America to the Northern Hemisphere (Europe and Asia), U.S. scientists broadened interests and cooperation with carbon science activities abroad. In 2007, the interagency working group initiated discussions with European Union (EU) officials under the framework of the U.S. Bilaterals and then in principle supported the early activities of European scientists leading the EU CarboEurope and CarboOcean programs on a proposed EU initiative to coordinate (network) data management activities on global carbon observations. The European Union and EU scientists aim to link European carbon observation research to similar existing programs in other countries as a contribution to the intergovernmental *ad hoc* Group on Earth Observations (GEO).

- *Global Partners* — The CCIWG initiated liaison proceedings with international nongovernmental organizations of the Earth System Science Partnership (ESSP). In 2006, the interagency working group established formal terms of reference with the ESSP Global Carbon Project (GCP), the GCP International Project Office, and the GCP Scientific Steering Committee (SSC). In 2007, the GCP SSC embraced the U.S. Carbon Cycle Science Program Office as their Affiliate Project Office in North America. The international link with the ESSP GCP is expected to yield stronger national program interactions, regional and global syntheses, and relevant deliverables on the global carbon budgets.

These activities will address Goals 2, 3, and 4
and Questions 7.4 and 7.5 of the CCSP Strategic Plan.

GLOBAL CARBON CYCLE
CHAPTER REFERENCES

1) **Raupach**, M.R., G. Marland, P. Ciais, C. Le Quere, J.G. Canadell, G. Klepper, and C.B. Field, 2007: Global and regional drivers of accelerating CO_2 emissions. *Proceedings of the National Academy of Sciences*, **104**, 10288-10293.

2) **Magnani**, F., M. Mencuccini, M. Borghetti, P. Berbigier, F. Berninger, S. Delzon, A. Grelle, P. Hari, P.G. Jarvis, P. Kolari, A.S. Kowalski, H. Lankreijer, B.E. Law, A. Lindroth, D. Loustau, J. Manca, J. Moncrieff, M. Rayment, V. Tedeschi, R. Valentini, and J. Grace, 2007: The human footprint in the carbon cycle of established temperate and boreal forests. *Nature*, **447**, 848-850.

3) **Irvine**, J., B.E. Law, and K. Hibbard, 2007: Post-fire carbon pools and fluxes in semi-arid ponderosa pine in Central Oregon. *Global Change Biology*, **13**, 1-13.

GLOBAL CARBON CYCLE
CHAPTER REFERENCES (CONTINUED)

4) **King**, J.S., C.P. Giardina, K.S. Pregitzer, and A.L. Friend, 2007: Biomass partitioning in red pine (*Pinus resinosa*) along a chronosequence in the Upper Peninsula of Michigan. *Canadian Journal of Forest Research*, **37**, 93-102.

5) **Parton**, W., W.L. Silver, I.C. Burke, L. Grassens, M.E. Harmon, W.S. Currie, J.Y. King, E.C. Adair, L.A. Brandt, S.C. Hart, and B. Fasth, 2007: Global-scale similarities in nitrogen release patterns during long-term decomposition. *Science*, **315**, 361-364.

6) **Bates**, N.R. and A.J. Peters, 2007: The contribution of acid deposition to ocean acidification in the subtropical North Atlantic Ocean. *Marine Chemistry*, **107**, 547-558.

7) **Doney**, S.C., N. Mahowald, I. Lima, R.A. Feely, F.T. Mackenzie, J.-F. Lamarque, and P.J. Rasch, 2007: The impact of anthropogenic atmospheric nitrogen and sulfur deposition on ocean acidification and the inorganic carbon system. *Proceeding of the National Academy of Sciences*, **104**, 14580-14585, doi:10.1073/pnas.0702218104.

8) **Fitch**, D.T. and J.K. Moore, 2007: Wind speed influence on phytoplankton bloom dynamics in the Southern Ocean Marginal Ice Zone. *Journal of Geophysical Research*, **112**, C08006, doi:10.1029/2006JC004061.

9) **Walvoord**, M.A. and R.G. Striegl, 2007: Increased groundwater to stream discharge from permafrost thawing in the Yukon River basin: Potential impacts on lateral export of carbon and nitrogen. *Geophysical Research Letters*, **34**, L12402, doi:10.1029/2007GL030216.

10) **Zhuang**, Q., J.M. Melillo, M.C. Sarofim, D.W. Kicklighter, A.D. McGuire, B.S. Felzer, A. Sokolov, R.G. Prinn, P.A. Steudler, and S. Hu, 2006: CO_2 and CH_4 exchanges between land ecosystems and the atmosphere in northern high latitudes over the 21st century. *Geophysical Research Letters*, **33**, L17403, doi:10.1029/2006GL026972.

11) **Zhuang**, Q., J.M. Melillo, A.D. McGuire, D.W. Kicklighter, R.G. Prinn, P.A. Steudler, B.S. Felzer, and S. Hu, 2007: Net emissions of CH_4 and CO_2 in Alaska: Implications for the region's greenhouse gas budget. *Ecological Applications*, **17**, 203-212.

12) **Balshi**, M.S., A.D. McGuire, Q. Zhuang, J.M. Melillo, D.W. Kicklighter, E.S. Kasischke, C. Wirth, M. Flannigan, J. Harden, J.S. Clein, T.J. Burnside, J. McAllister, W.A. Kurz, M. Apps, and A. Shvidenko, 2007: The role of fire disturbance in the carbon dynamics of the pan-boreal region: A process-based analysis. *Journal of Geophysical Research*, **112**, G02029, doi:10.1029/2006JG000380.

13) **Gu**, L., W.M. Post, D. Baldocchi, T.A. Black, A.E. Suyker, S.B. Verma, T. Vesala, and J.W. Munger, 2008: Characterizing the seasonal dynamics of plant community photosynthesis across a range of vegetation types. In: *Phenology of Ecosystem Processes* [Noormets, A. and L. Gu (eds.)]. Springer Science + Business Media, New York, NY, USA, in press.

14) **Saleska**, S.R., K. Didan, A.R. Huete, and H.R. da Rocha, 2007: Amazon forests green-up during 2005 drought. *Science*, **318**, 612, doi:10.1126/science.1146663.

15) **Daughtry**, C.S.T., P.C. Doraiswamy, E.R. Hunt Jr., A.J. Stern, J.E. McMurtrey III, and J.H. Prueger, 2006: Assessing crop residue cover and soil tillage intensity. *Journal of Soil and Tillage Research*, **91**, 101-108.

16) **Doraiswamy**, P.C., G.M. McCarty, E.R. Hunt, R. Yost, M. Doumbia, and A.J. Franzluebbers, 2007: Modeling of soil carbon sequestration in agricultural lands of Mali. *Agricultural Systems*, **94(1)**, 63-74.

17) **Tilman**, D., J. Hill, and C. Leyman, 2006: Carbon-negative biofuels from low-input high-diversity grassland biomass. *Science*, **314**, 1598-1600.

18) **CCSP**, 2007: *The First State of the Carbon Cycle Report (SOCCR): The North American Carbon Budget and Implications for the Global Carbon Cycle.* A Report by the U.S. Climate Change Science Program and the Subcommittee on Global Change Research [King, A.W., L. Dilling, G.P. Zimmerman, D.M. Fairman, R.A. Houghton, G. Marland, A.Z. Rose, and T.J. Wilbanks (eds.)]. National Oceanic and Atmospheric Administration, National Climatic Data Center, Asheville, NC, USA, 242 pp.

19) **Peters**, W., A.R. Jacobson, C. Sweeney, A.E. Andrews, T.J. Conway, K. Masarie, J.B. Miller, L.M.P. Bruhwiler, G. Pétron, A.I. Hirsch, D.E.J. Worthy, G. van der Werf, J.T. Randerson, P.O. Wennberg, M.C. Krol, and P.P. Tans, 2007: An atmospheric perspective on North American carbon dioxide exchange: CarbonTracker. *Proceedings of the National Academy of Sciences*, **104**, 18925-18930.

6 | Ecosystems

Strategic Research Questions

8.1 What are the most important feedbacks between ecological systems and global change (especially climate), and what are their quantitative relationships?

8.2 What are the potential consequences of global change for ecological systems?

8.3. What are the options for sustaining and improving ecological systems and related goods and services, given projected global changes?

See Chapter 8 of the *Strategic Plan for the U.S. Climate Change Science Program* for detailed discussion of these research questions.

The terrestrial and aquatic ecosystems that make up the biosphere provide vital goods and services to humanity, including food, fiber, fuel, genetic resources, pharmaceuticals, cycling and purification of water and air, regulation of weather and climate, recreation, and natural beauty. Recent and ongoing global environmental changes—including climatic change, changes in atmospheric composition, land-use change, habitat fragmentation, pollution, and the spread of invasive species—are affecting the structure, composition, and functioning of many ecosystems, and therefore the goods and services that they provide. In turn, many ecological effects of global environmental change have potential to affect atmospheric composition, weather, and climate through both negative and positive feedback mechanisms. Because many global environmental changes are expected to increase in magnitude in the coming decades, the potential exists for increased effects of climate change on ecosystems and the goods and services that they provide (see Figure 16). Improved understanding of potential effects of global change on ecosystems, as well as the feedbacks from ecosystems to global change processes, remains a CCSP priority.

In FY 2009, the CCSP Ecosystems Interagency Working Group (EIWG) will continue with its planning, implementation, and analysis of research programs to accomplish the CCSP Strategic Plan goals related to ecosystem research. One focus will be increased efforts to provide the scientific basis needed for improved forecasts of the effects of climatic change on the structure, composition, and functioning of terrestrial and aquatic ecosystems, including the many goods and services that they provide. EIWG will also continue its FY 2007 and FY 2008 focus on the interplay between changing climate and the productivity and biodiversity of ecosystems, with an emphasis on improving understanding of ecological processes to accelerate model development and analysis. This research will include funding for two topics of particular urgency: (1) vulnerability of coastal ecosystems (both terrestrial and aquatic) to oceanic warming, sea-level rise, increased storm frequency or intensity, saltwater intrusion, and increased sedimentation and runoff; and (2) warming-induced changes in high-latitude and high-elevation ecosystems, including changes in primary production, species composition, the timing of water availability, and migration of the tree line and other ecotones. These topics require additional research on underlying ecological processes and responses and the development of models linking geophysical and ecological phenomena. Strategies for implementation include new *in situ* experimental research projects; observations of ecosystems at local, regional, and global scales; synthesis and analysis of diverse ecological data sets, including those from manipulative experiments; and ecological model development and evaluation.

EIWG efforts contribute to all five CCSP goals, with an emphasis on Goal 4: to "understand the sensitivity and adaptability of different natural and managed ecosystems…to climate and related global changes." EIWG activities directly address questions 8.1, 8.2, and 8.3 from the CCSP Strategic Plan. Synergies and interactions exist with all the other CCSP research elements (i.e., Atmospheric Composition, Climate Variability and Change, Global Carbon Cycle, Global Water Cycle, Land-Use and Land-Cover Change, and Human Contributions and Responses).

The agencies participating in the EIWG work collaboratively to plan and execute research described in the CCSP Strategic Plan. Many of the research accomplishments and plans described in this chapter are the outcome of multi-agency efforts. A number of these activities also involve collaborations between the agencies and non-Federal partners and cooperators. EIWG actively engages the larger scientific research community to obtain input to and feedback on its evolving research plans.

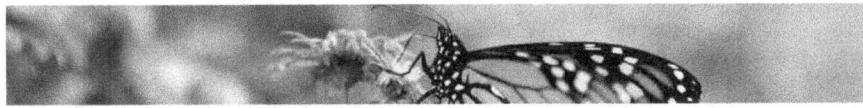

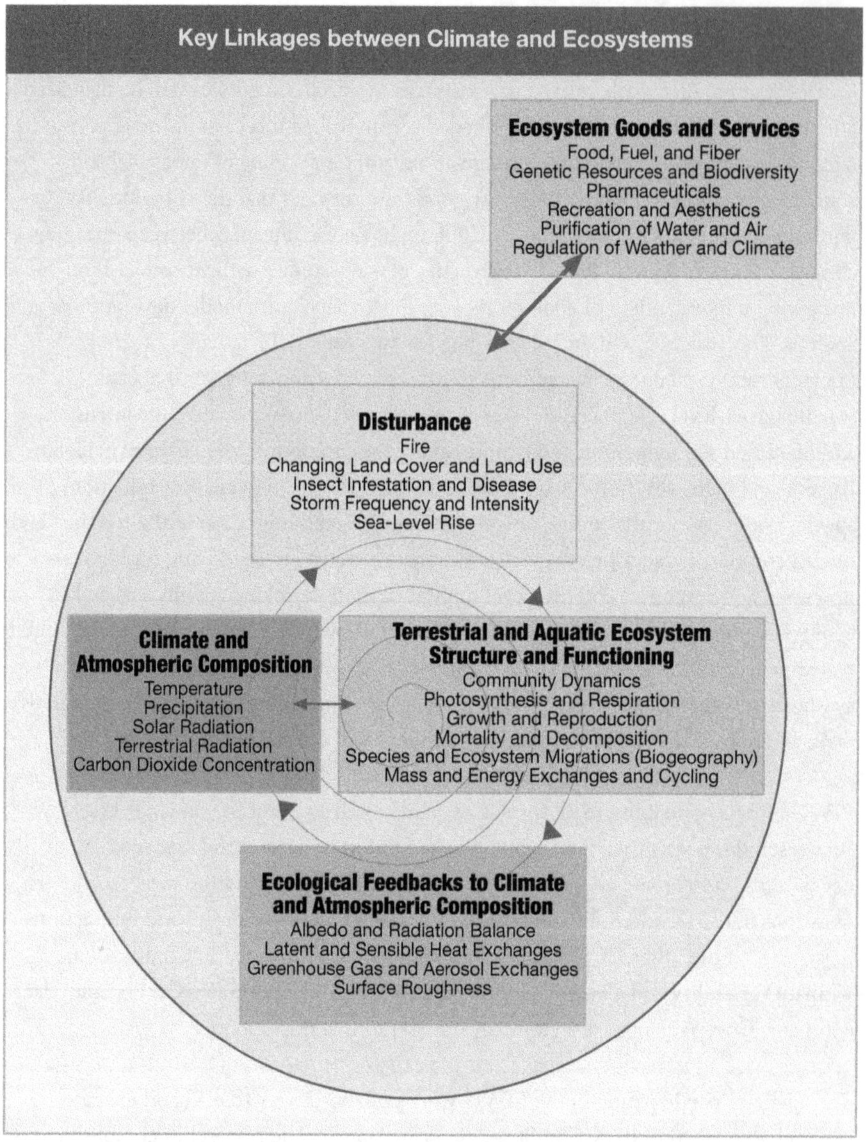

Figure 16: Key Linkages between Climate and Ecosystems. Climate changes (i.e., changes in temperature, precipitation, CO$_2$ concentration, wind, or solar or terrestrial radiation) can affect terrestrial and aquatic ecosystems by altering primary production processes, reproduction, health and mortality of organisms, and rates and pathways of decomposition, community dynamics and biogeography, and exchanges of mass and energy between ecosystems and the atmosphere. Climate changes also have the potential to affect the frequency and magnitude of various ecosystem disturbances (e.g., fire, disease, insect infestations, storm frequency, and land-use change). In turn, changes in ecosystem-atmosphere exchanges of radiation, heat, or greenhouse gases caused directly or indirectly by climate change have the potential to dampen or enhance the initial climatic change through negative or positive feedbacks. Ecosystem changes caused by climatic changes can also affect the many ecosystem goods and services on which society depends. Likewise, climate change effects on ecosystem goods and services may elicit human actions that in turn affect climate, ecosystem disturbance, and/or ecosystem structure and functioning. Temporal and spatial scales are implicit; temporal scales range from seconds to millennia and spatial scales range from local to global. *Credit: CCSP Ecosystems Interagency Working Group.*

HIGHLIGHTS OF RECENT RESEARCH

Climate Interactions with Tree Mortality and Regeneration are Complex.[1,2] **Recent research is** improving understanding of the complex interactions that can determine the effect of changing climate on vegetation mortality and regeneration. In the southern Appalachians, factors favoring seedling regeneration of particular tree species include warmer spring temperatures and dryer soil conditions, reduction of pathogen attacks on seeds in dryer soils, and warmer average temperatures. A synthesis of recent research indicates that with current climate trends, only higher elevation forests in this region are in danger of extinction. Changes in snowpack might be an important factor in other regions. Yellow cedar has been mysteriously dying in Alaska for over a century. After 20 years of study, researchers now hypothesize that this trend primarily results from decreasing snowpack, which increases soil freezing, kills vulnerable roots, and leads to foliar mortality and death. Recent tests of seedling survival in artificial snowpacks also support this mechanism as a factor in declining regeneration.

Soil Responses to Environmental Change.[3,4,5] **Experimental warming of tundra indicated** that ecosystem respiration increases most in dry tundra; that saturated soils dampen responses in wetter tundra; and that both carbon gain and respiration respond to warming. A Florida scrub oak ecosystem exposed to elevated carbon dioxide (CO_2)

Florida Scrub Experiment

showed the largest increase in plant growth reported to date, but a 25% decline in soil carbon offset half of this increase in carbon accumulation, indicating that models may overestimate the potential for ecosystems to slow atmospheric CO_2 increase by storing carbon in soils (see Figure 17). Finally, in a constructed old-field ecosystem in eastern Tennessee, soil respiration increased with elevated atmospheric CO_2 and decreased with reduced soil moisture, but responses to

Figure 17: Florida Scrub Experiment.
A soil core from the Florida scrub experiment. *Credit: B. Hungate, Northern Arizona University.*

experimental warming varied. This complexity in responses indicates that models should include mechanistic representations of the effects of multiple global change factors on ecosystem processes.

Competitive Interaction of Trees Altered by Carbon Dioxide and Ozone.[6] Changes in atmospheric CO_2 and ozone (O_3) concentrations may alter competitive interactions between tree species. Results from a long-term (10-year) field study in northern Wisconsin indicate that rising atmospheric CO_2 or O_3 concentrations would give a competitive advantage to paper birch over trembling aspen. Birch was less sensitive than aspen to damage by O_3, and in mixed stands birch was more responsive to the positive effects of rising CO_2 than aspen. While earlier, shorter duration experiments found little influence of these two gases on forest succession, these new results indicate that rising concentrations of either CO_2 or O_3 could alter the species composition of mixed aspen-birch stands that now cover millions of acres in the north central and northeastern United States.

Changes in Habitat Suitability for Tree and Bird Species.[7,8] One of the first steps in projecting possible impacts of climate change on plant and animal distributions and developing adaptive approaches to forest management is understanding how the distribution of environmental conditions suitable for individual species is likely to change over time. A new Climate Change Atlas web site presents results of extensive modeling efforts that combine current environmental relationships and species distributions with global climate models to project changes in potential suitable habitat for 134 trees and 150 birds of eastern North America by the end of the century (see Figure 18). Each species was modeled individually to show current distribution and potential distribution of suitable habitat in the future according to regionalized outputs of models for two Intergovernmental Panel on Climate Change (IPCC) emission scenarios and three climate models.

Modeling Impacts of Changing Climate on Intertidal Species Distributions.[9] A rich historical data record makes the intertidal zone a model system for examining the effects of climate change. Comparisons of historical and 2006 geographic distributions of the arctic barnacle *Semibalanus* and the tropical sand-worm *Diopatra* show parallel northward shifts at rates of 15 to 50 km per decade since 1872. Using modeled climate data from weather reanalyses, which provide the large-scale environmental envelope in which organisms existed over time, and simulation models of animal body temperatures, which allow simulation of organism response to the environment, researchers accurately modeled changes in the distributions of the two species over the past 50 years. Parallel shifts in distribution indicate that similar responses to a warming climate control the geographic limits of both species.

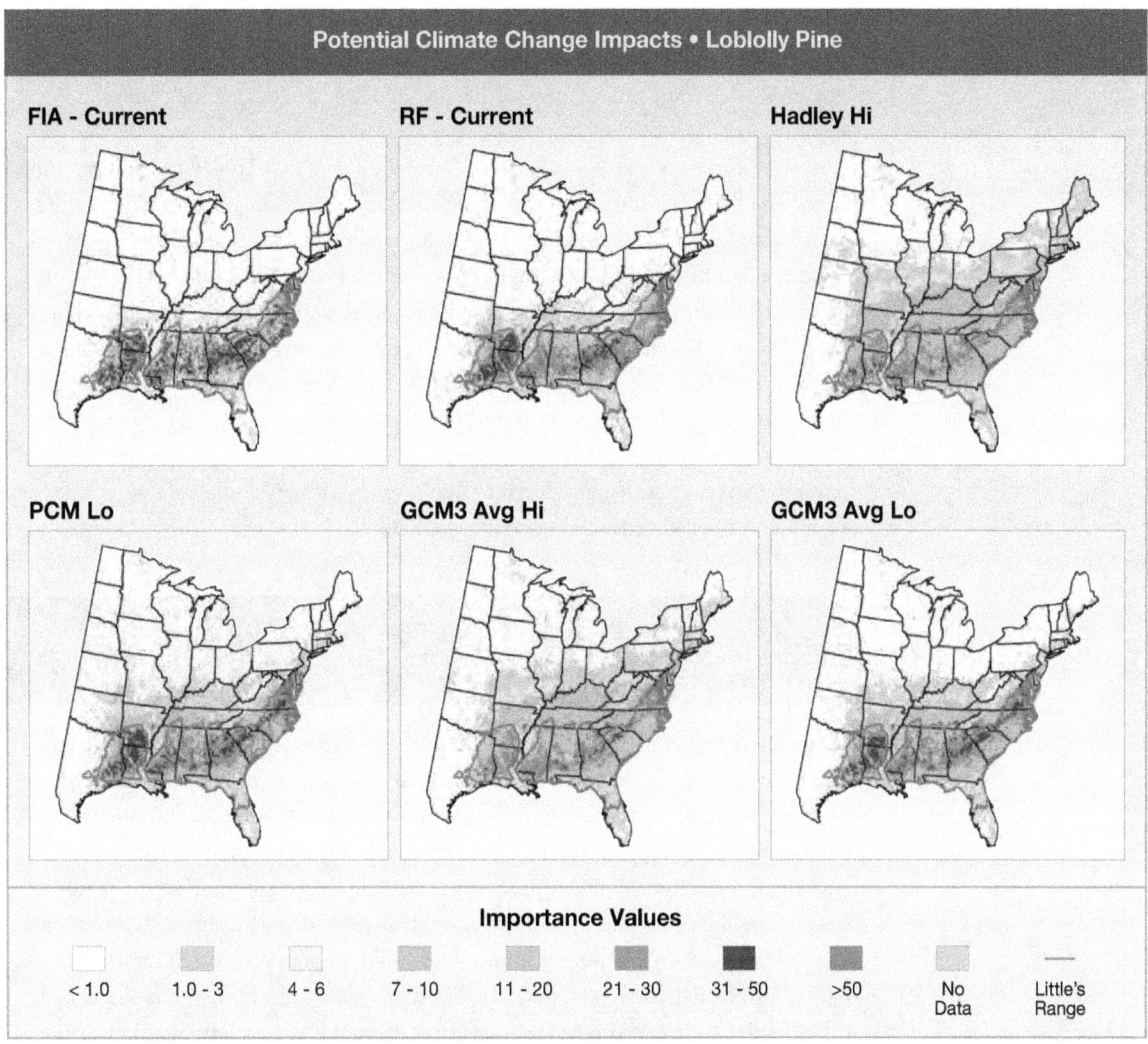

Figure 18: Potential Climate Change Impacts • Loblolly Pine. An example summary output from the Climate Change Tree Atlas (<nrs.fs.fed.us/atlas>) illustrates the potential impacts of changing climate on habitat suitability for loblolly pine under different climate change scenarios relative to current mapped (upper left) and modeled (upper center) distributions. The remaining four panels illustrate average projections from two general circulation models [Hadley CM3, DOE Parallel Climate Model (PCM)] as well as average projections (GCM3 Avg) from three models (Hadley, PCM, and Geophysical Fluid Dynamics models) under high- and low-emission scenarios. *Credit: Adapted from A.M.Prasad, L.R. Iverson, S. Matthews, and M. Peters, USDA Forest Service.*

Climate Impacts on Zooplankton Composition and Fish Feeding Preferences.[10] A series of studies on the Pribilof Island ecosystem in the eastern Bering Sea investigated why this area is able to support a large biomass of top predators, and whether the mechanisms responsible for production are sensitive to climate variability. One of these studies documented differences in the zooplankton community and the food ingested by juvenile walleye pollock on the Bering Sea shelf and in the Pribilof ecosystem for relatively cold and warm climate conditions. During the warm year, there were significantly fewer large, lipid-rich zooplankton prey for the juvenile fish than during the cool year.

Climate and Marine Fisheries—Spatial and Temporal Variability.[11,12] **Understanding** variation in time and space is critical to evaluating impacts of ocean changes on marine resources. Information from marked hatchery salmon has revealed local covariability in survival between adjacent coho stocks within a region of the Pacific Ocean and coherence in survival in adjacent regions, but no clear evidence of covariability at greater spatial scales. Other research suggests that Labrador Sea seawater from melting of the Greenland Ice Sheet is causing freshening of surface waters in the Gulf of Maine and Georges Bank. This fresher water flows southwest along the coast, keeping plankton near the surface under better growing conditions. Researchers expect higher plankton concentrations to provide more food for fish larvae, with higher survival and recruitment of adult cod and haddock stocks on Georges Bank.

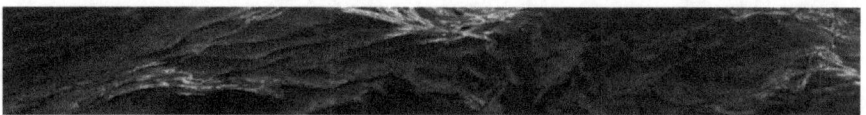

California Current: Disruptions in Seasonal Cycles of Production.[13,14,15,16] **California** Current ecosystems are characterized by strong seasonal variability in productivity driven by the strength and duration of upwelling. Climate change scenarios suggest disruptions in phenology and biological interactions that depend upon seasonally predictable upwelling cycles. Recent observations support these expectations: A 4-year period of strong upwelling, cold ocean conditions, and high productivity from 1999 to 2002 was followed by 4 years of delayed or reduced upwelling, warm ocean conditions, and reduced productivity. In 2005, due to delayed upwelling, plankton biomass declined, seabirds failed to fledge young, and survival of salmon was very low. Should warm ocean conditions continue (as from 2003 to 2006), productivity in the California Current may decline.

Identifying Factors Contributing to Coral Reef Resilience.[17,18] **Climate** variability and change can negatively affect coral reef ecosystems (see Figure 19). Effective management needs to assess reef vulnerabilities, identify adaptive management strategies, and integrate these with existing decisions and mandates. Results from American Samoa show that adaptive management strategies could be implemented to increase reef resilience. A second study demonstrates how marine reserves contribute to resilience. Caribbean reefs became susceptible to changing from one stable state to another after a sea urchin (*Diadema antillarum*) die-off. Although the establishment of marine reserves increased

116

Partially Bleached Pacific Coral Reef

Figure 19: Partially Bleached Pacific Coral Reef. Coral reef species and communities vary in their resilience (ability to resist or recover) in the face of climate change impacts such as coral bleaching. *Credit: E. Mielbrecht, Emerald Coast Environmental Consulting.*

predation on parrotfishes, the current dominant species, protection from overfishing had a greater impact and ultimately allowed parrotfish densities to further increase. Increased grazing pressure from parrotfishes caused a fourfold reduction in cover of macroalgae, the principal competitors of corals.

Highlights of Plans for FY 2009

Establishing a National Phenological Network. Phenological data and models are important to agriculture, drought monitoring, and wildfire risk assessment, as well as management of invasive and pest species and infectious diseases. Existing phenological records are largely short term and spotty, however. In response, the U.S. National Phenological Network (NPN) is an emerging partnership among the academic community, several Federal agencies, and volunteers designed to fill that void through integration of data, analysis, and modeling. NPN will create four products: a meta-database of existing phenological data, a set of data collection and management protocols, lists of target species and infectious diseases representative of ecoregions, and an enhanced web site to facilitate communication and data access. While phenological models are important to infectious diseases and may be useful, few such models currently exist. Members of the public will be substantially involved as citizen-scientists in NPN.

This activity will address Questions 8.1 and 8.2 of the CCSP Strategic Plan.

Linking Arctic Climate Change, Societies, and Natural Systems. High-latitude ecosystems are experiencing the most pronounced environmental changes on Earth, making new research initiated as part of the fourth International Polar Year (IPY, 2007-2009) especially timely. Research will examine how key human services provided by natural ecosystems are changing throughout Alaska as the dynamic linkages between human social systems (both native and non-native) and natural systems change as warming continues to occur over time. Of particular focus in the study will be walrus, caribou, moose, wild berries, and forest fires.

This activity will address Questions 8.1, 8.2, and 8.3 of the CCSP Strategic Plan.

Experimental Study of Warming at the Alpine Tree Line. Warming of several degrees Celsius, as projected for this century by climate models, has the potential to force the alpine tree line above the peak elevation of many western mountains of the United States. Such an upward "migration" of the alpine tree line might displace or eliminate existing alpine ecosystems (e.g., alpine meadows now found above the tree line). A set of experimental manipulations of temperature at the alpine tree line in the western United States is planned to begin in FY 2009. The experiments will determine effects of increased temperature on the ability of trees to germinate and grow at elevations above the present tree line, and thus to potentially displace existing alpine ecosystems if they experience future warming.

This activity will address Question 8.2 of the CCSP Strategic Plan.

Potential Impacts of Climate Change on Wildlife Habitats. Impacts of climate on vegetation are expected to significantly affect wildlife habitat and diversity. Wildlife agencies need information on these impacts as well as potential options for ameliorating them. Scientists will synthesize information on interactions between climate and vegetation change to quantify potential effects of predicted changes in habitat on terrestrial vertebrate biodiversity and identify management options.

This activity will address Questions 8.2 and 8.3 of the CCSP Strategic Plan.

Tree Growth and Stand Dynamics Projections incorporating Climate Change. Scientists and managers will develop and evaluate approaches for integrating climate change into the Forest Vegetation Simulator. The simulator is widely used by forest planners and fire managers in the Forest Service, DOI, and other agencies to project effects on tree growth and stand-level

dynamics of thinning, prescribed fire, and other treatments, as well as effects of insect and disease interactions.

This activity will address Question 8.3 of the CCSP Strategic Plan.

Understanding Tropical Diversity with Satellite, Morphological, and Molecular Data. A combination of satellite remote sensing, measures of morphological traits from target bird species, and genetic markers is shedding light on the drivers of species diversification in the Ecuadorian Andes, a neotropical biodiversity hotspot. Morphological data allow detection of traits under selection and genetic techniques reveal the role of geographic barriers (e.g., mountains and rivers) in promoting divergent evolution. Remote-sensing data allow correlation of patterns of variation in the biological information obtained on the ground with climatic, topographical, and other environmental variables (e.g., vegetation characteristics) in order to elucidate some of the factors driving diversification. Correlating environmental landscapes with "morphological landscapes" and "genetic landscapes" will allow mapping of both biodiversity patterns and the underlying processes, vital information for projecting effects of climate change.

This activity will address Question 8.2 of the CCSP Strategic Plan.

Nonlinear Responses to Global Change in Aquatic Ecosystems. Several research efforts focus on identifying nonlinear responses to global change in aquatic ecosystems. Potential "regime shifts" involve the fundamental reorganization of natural ecosystems as environmental conditions change. They are difficult to predict, but may occur rapidly, and have potentially large consequences for ecosystem services. A new manipulative experiment will test predictions relating the stability of a Michigan lake to the structure of its food web. If predictions are correct, then it should become increasingly possible to forecast such changes in other ecosystems, perhaps in time to intervene. New modeling efforts will also contribute to the prediction of nonlinear responses. These results may be incorporated into local planning and management processes, to help prevent regime shifts in other aquatic ecosystems

This activity will address Questions 8.2 and 8.3 of the CCSP Strategic Plan.

Implications of Climate Change for Biological Indicators and Invasive Species. Research will be conducted on how biological indicators may be used to detect or control climate

change effects in aquatic ecosystems, including changes in community composition, phenology, reproductive rate, evolutionary adaptations, and genetic selection. Assessment programs relying on biological indicators to document ecosystem condition will use this research to identify climate effects on ecosystems. This information allows modification of assessment programs to account for effects and to ensure that management goals continue to be met. Another indicator of potential changes is aquatic invasive species. Research will focus on implications of climate change effects on the invasion pathway and ecosystem management. A synthesis of management activities; effects on aquatic organisms, pathways, and ecosystem services; and available literature will describe adaptation options for aquatic invasive species management.

This activity will address Questions 8.2 and 8.3 of the CCSP Strategic Plan.

Impacts of Climate Change on Marine Fisheries. **Regional ecosystem studies will be** conducted to develop and test assessment and prediction capabilities to aid fisheries management. Monitoring and process studies will help determine the mechanisms and rates of climate impacts on ecosystems. Studies will also synthesize historical information on species of interest and collect new data necessary to construct models of their larval transport. The models will eventually simulate transport scenarios for regional changes in wind and runoff patterns, which drive the ocean currents. Reproductive success of many important fisheries species depends upon the transport of their eggs and larvae to suitable habitat. The development and testing of recruitment and environmental indices will also continue in order to increase their accuracy and precision for predicting important ecosystem changes.

This activity will address Question 8.2 of the CCSP Strategic Plan.

Ocean Acidification—Changing Oceans, Changing Ecosystems. **Rising atmospheric** CO_2 levels are altering ocean chemistry and threatening marine biodiversity. Decreasing oceanic pH resulting from increasing atmospheric CO_2 reduces the abilities of calcifying organisms, such as corals and crustaceans, to form skeletons and shells. It is increasingly urgent to have a mechanistic understanding of marine carbonate chemistry, including historical fluctuations and current trends, as well as predicting the responses of marine ecosystems to increased acidity (reduced pH). Upcoming research will include studies on historical fluctuations of pH based on the geological record, monitoring and establishing long-term time series of marine carbonate chemistry, *in situ* monitoring of calcification rates of key species, evolutionary change in organisms in response to changing chemistry, compensatory shifts in species within functional groups and the role of biodiversity in facilitating such shifts, effects of lower pH on coral calcification, functional genomics studies of pH effects on molecular regulation of calcification, and development of models to predict effects of multiple environmental changes on organism, population, and ecosystem-level adaptation.

This activity will address Question 8.2. of the CCSP Strategic Plan.

ECOSYSTEMS CHAPTER REFERENCES

1) **Ibanez**, I., J.S. Clark, S. LaDeau, and J.H. Ris Lambers, 2007: Exploiting variability to understand tree recruitment response to climate change. *Ecological Monographs*, **77**, 163-177.

2) **Hennon**, P., D. D'Amore, D. Wittwer., A. Johnson., P. Schaberg, G. Hawley, C. Beier, S. Sink, and G. Juday, 2006: Climate warming, reduced snow, and freezing injury could explain the demise of yellow cedar in southeast Alaska. *World Resource Review*, **18**, 427-445.

3) **Oberbauer**, S.F., C.E. Tweedie, J.M. Welker, J.T. Fahnestock, G.H.R. Henry, P.J. Webber, R.D. Hollister, M.D. Walker, A. Kuchy, E. Elmore, and G. Starr, 2007: Tundra CO_2 fluxes in response to experimental warming across latitudinal and moisture gradients. *Ecological Monographs*, **77**, 221-238.

4) **Carney**, K.M., B.A. Hungate, B.G. Drake, and J.P. Megonigal, 2007: Altered soil microbial community at elevated CO2 leads to loss of soil carbon. *Proceedings of the National Academy of Sciences*, **104**, 4990-4995.

5) **Wan**, S.Q., R.J. Norby, J. Ledford, and J.F. Weltzin, 2007: Responses of soil respiration to elevated CO_2, air warming, and changing soil water availability in a model old-field grassland. *Global Change Biology*, **13**, 2411-2424, doi:10.1111/j.1365-2486.2007.01433.x.

6) **Kubiske**, M.E., V.S. Quinn, P.E. Marquardt, and D.F. Karnosky, 2007: Effects of elevated CO_2 and/or O_3 on intra- and interspecific competitive ability of aspen. *Plant Biology*, **9**, 342-355.

7) **Prasad**, A.M., L.R. Iverson, S. Matthews, and M. Peters, 2007-ongoing: *A Climate Change Atlas for 134 Forest Tree Species of the Eastern United States* [database]. Northern Research Station, USDA Forest Service, Delaware, Ohio. <www.nrs.fs.fed.us/atlas/tree>.

8) **Iverson**, L.R., A.M. Prasad, S.N. Matthews, and M. Peters, 2008: Estimating potential suitable habitat for 134 eastern US tree species under six climate scenarios. *Forest Ecology and Management*, **254**, 390-406, doi:10.1016/j.foreco.2007.07.023.

9) **Wethey**, D.S. and S.A. Woodin, 2008: Ecological hindcasting of biogeographic responses to climate change in the European intertidal zone. *Hydrobiologia*, **606**, 139-151, doi:10.1007/s10750-008-9338-8.

10) **Coyle**, K.O., A.I. Pinchuk, L. Eisner, and J.M. Napp, 2008: Zooplankton species composition, abundance, and biomass on the southeastern Bering Sea shelf during summer: the potential role of water column stability and nutrients in structuring the zooplankton community. *Deep-Sea Research II, Topical Studies in Oceanography*. in press.

11) **Teo**, S.L.H., L.W. Botsford, and A. Hastings, 2008. Spatio-temporal covariability in coho salmon (Oncorhynchus kisutch) survival, from California to Southeast Alaska. *Deep-Sea Research II, Topical Studies in Oceanography*. submitted.

12) **Greene**, C.H. and A.J. Pershing, 2007: Climate drives sea change. *Science*, **315**, 1084-1085.

13) **Barth**, J.A., B.A. Menge, J. Lubchenco, F. Chan, J.M. Bane, A.B. Kirincich, M.A. McManus, K.J. Nielsen, S.D. Pierce, and L. Washburn, 2007: Delayed upwelling alters nearshore coastal ocean ecosystems in the northern California Current. *Proceedings of the National Academy of Science*, **104**, 3719-3724.

14) **Mackas**, D.L., W.T. Peterson, M.D. Ohman, and B.E. Lavaniegos, 2006: Zooplankton anomalies in the California Current system before and during the warm ocean conditions of 2005. *Geophysical Research Letters*, **33**, L22S07, doi:10.1029/2006GL027930.

15) **Pierce**, S.D., J.A. Barth, R.E. Thomas, and G.W. Fleischer, 2006: Anomalously warm July 2005 in the northern California Current: historical context and the significance of cumulative wind stress. *Geophysical Research Letters*, **33**, L22S04, doi:10.1029/2006GL027149.

16) **Sydeman**, W.J., R.W. Bradley, P. Warzybok, C.L. Abraham, J. Jahncke, K.D. Hyrenbach, V. Kousky, J.M. Hipfner, and M.D. Ohman, 2006: Planktivorous auklet *Ptychoramphus aleuticus* responses to ocean climate, 2005: unusual atmospheric blocking? *Geophysical Research Letters*, **33**, L22S09, doi:10.1029/2006GL026736.

17) **USEPA**, 2007: *Climate Change and Interacting Stressors: Implications for Coral Reef Management in American Samoa*. EPA/600/R-07/069. Global Change Research Program, National Center for Environmental Assessment, Washington, DC.

18) **Mumby**, P.J., C.P. Dahlgren, A.R. Harborne, C.V. Kappel, F. Micheli, D.R. Brumbaugh, K.E. Holmes, J.M. Mendes, K. Broad, J.N. Sanchirico, K. Buch, S. Box, R.W. Stoffle, and A.B. Gill, 2006: Fishing, trophic cascades, and the process of grazing on coral reefs. *Science*, **311**, 98-101.

7 | Human Contributions and Responses and Decision-Support Resources Development

Human Contributions and Responses to Environmental Change Strategic Research Questions

9.1   What are the magnitudes, interrelationships, and significance of the primary human drivers of, and their potential impact on, global environmental change?

9.2   What are the current and potential future impacts of global environmental variability and change on human welfare, what factors influence the capacity of human societies to respond to change, and how can resilience be increased and vulnerability reduced?

9.3   How can the methods and capabilities for societal decisionmaking under conditions of complexity and uncertainty about global environmental variability and change be enhanced?

9.4   What are the potential human health effects of global environmental change, and what climate, socioeconomic, and environmental information is needed to assess the cumulative risk to health from these effects?

The Role of Decision-Support Resources Development

Goal 1:   Prepare scientific syntheses and assessments to support informed discussion of climate variability and change and associated issues by decisionmakers, stakeholders, the media, and the general public.

Goal 2:   Develop resources to support adaptive management and planning for responding to climate variability and climate change, and transition these resources from research to operational application.

Goal 3:   Develop and evaluate methods (scenario evaluations, integrated analyses, and alternative analytical approaches) to support climate change policymaking and demonstrate these methods with case studies.

See Chapter 9 of the *Strategic Plan for the U.S. Climate Change Science Program* for detailed discussion of the strategic research questions and Chapter 11 for decision-support resources development.

Human activities play an important role in the Earth system and are significant drivers
of change in the environment at all scales—local, regional, national and global. Humans
also have the capability to respond to changes in their environment, and adaptations,
when effective, enhance the resilience of both managed and natural systems. At the same
time, social and economic systems are changing in a world that is more interconnected
than ever. A better integrated understanding of the complex interactions between human
societies and the Earth system is needed to identify vulnerable systems and pursue
options that take advantage of opportunities to enhance resilience. The National Research
Council (NRC) report, *Climate Change Science: An Analysis of Some Key Questions,*[1] concluded
that: "In order to address the consequences of climate change and better serve the
Nation's decisionmakers, the research enterprise dealing with environmental change
and environment-society interactions must be enhanced." Such an enterprise should
include "…support of interdisciplinary research that couples physical, chemical,
biological, and human systems."

The study of the human interface with change in the global environment is especially
important because of its capacity to inform public policy. Decisionmaking, however, is
challenged by uncertainties, including risks of irreversible and/or nonlinear changes
that may be met with insufficient or excessive responses whose consequences may
cascade across generations. The difficulties associated with uncertainty have become
increasingly salient given the interest of policymakers in addressing global environmental
change. Uncertainties arise from a number of factors, including problems with data,
problems with models, lack of knowledge of important underlying relationships,
imprecise representation of uncertainty, statistical variation and measurement error,
and subjective judgment.[2]

It is well established that human health is linked to environmental conditions, and
that changes in the natural environment may have subtle, or dramatic, effects on

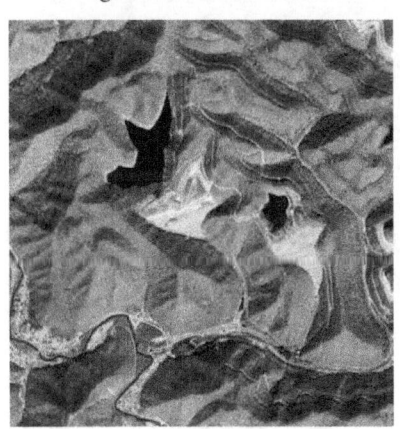

health. Timely knowledge of these effects
may support public health systems in
devising and implementing strategies to
compensate or respond to them. Federally
supported research has thus far provided
information on a broad range of health
effects of global change, including the
adverse effects of ozone, atmospheric
particles and aeroallergens, ultraviolet
(UV) radiation, vector- and water-borne
diseases, and heat-related illnesses and
mortality.

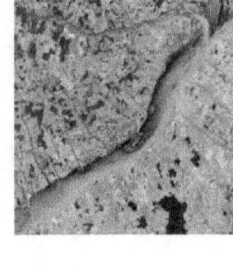

HIGHLIGHTS OF RECENT RESEARCH IN HUMAN CONTRIBUTIONS AND RESPONSES TO ENVIRONMENTAL CHANGE

Selected highlights of recent research and activities supported by CCSP-participating agencies follow.

Innovative Approaches to Next-Generation Scenario Development. The international climate modeling and integrated assessment modeling communities have begun planning research that could ultimately be reported in the Intergovernmental Panel on Climate Change (IPCC) Fifth Assessment Report. Researchers from these communities are coordinating research programs to deliver scenarios of potential future concentrations of a broad range of greenhouse gases and short-lived species of reactive gases and aerosols as well as land cover to be employed by the climate modeling community to produce ensemble climate calculations. The approach being developed requires that integrated assessment models produce representative concentration pathways (RCPs) that extend further into the future (to the year 2300) and contain greater detail (0.5° x 0.5°grid size) than any in the past. Work by the climate modeling community to incorporate the normally occurring processes of the carbon cycle means that integrated assessment models must deliver RCPs that include geographically disaggregated land-cover data as well as the greenhouse gases and short-lived species of reactive gases and aerosols. U.S. researchers are leading these efforts.

Energy and Environmental Impacts Research. In an effort to strengthen connections between integrated assessment research and impact and adaptation research, and in response to some of the identified challenges in CCSP Synthesis and Assessment Products (SAPs), a modest, focused research effort was initiated at the Oak Ridge National Laboratory (ORNL) to improve understanding of select energy and environmental impacts of climate change. For example, in FY 2007, CCSP completed the first comprehensive summary of the effects of climate change on energy production and use in the United States (SAP 4.5), and it cooperated with other CCSP agencies in addressing impact concerns ranging from cities to the carbon cycle. In a collaboration between ORNL, the Joint Global Change Research Institute (JGCRI), and the Massachusetts Institute of Technology (MIT), the findings are now being connected with integrated assessment tools and analyses, and this work is being extended by ORNL to the development of databases on U.S. and international impact research and science and on adaptation processes and managed practices. Additionally, a solicitation for FY 2009 research grants addressing these and other topics was published to engage the university community and

others in developing innovative approaches and methodologies for exploring and representing these impacts in integrated assessment models.

Connecting Communities of Practice: Integrated Assessment Research; Earth Systems Modeling; and Impacts, Adaptation, and Vulnerability (IAV). There has been substantial interagency cooperation over the past year in strengthening the capacity of the science community to incorporate an improved understanding of impacts and adaptation strategies in integrated assessments and in integrated modeling of the combined human and natural systems behaviors—both in driving climate change and responding to climate change. During the past year, a major week-long session was held in Snowmass, Colorado, which brought together ecologists, Earth system modelers, economists, energy experts, health experts, social scientists, specialists in adaptation, and multiple agency representatives, all of whom had been important contributors to the recently published IPCC reports. The exchange of perspectives, review of IPCC results, and presentation of more recent research has resulted in new collaborations being formed to investigate such issues as a more sophisticated integration of Earth system, IAV, and integrated assessment models, and an exploration of how the different research communities approach problems such as land use and land cover, with the intent of developing more coordinated approaches.

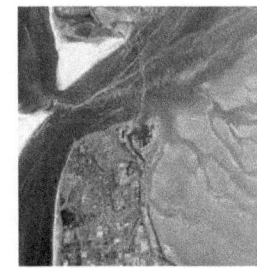

Interagency Scientific Workshops on Impacts and Adaptation in Integrated Assessment Research. Historically, the integrated assessment modeling community has focused on understanding the processes that lead to changes in emissions of greenhouse gases, aerosols, and other chemically active gases over time. It is now clear that understanding the physical and socioeconomic processes of impacts and adaptation are equally important in building a more complete understanding of global change. Over the past year, the integrated assessment research community, led by the JGCRI and ORNL, has convened a series of workshops and discussion sessions among agencies and active researchers in these fields to explore the major research challenges for integrated modeling and improved process understanding for adaptation and impacts. Workshop presentations and a synthesis report on research challenges are in preparation.

Workshop on Climate Change and Water. In October 2007, the Centers for Disease Control and Prevention (CDC) hosted a workshop entitled *Climate Change, Drinking Water, and Public Health*. The purpose of the workshop was to share information across various sectors, including environmental, public health, and water utilities, to extend the breadth and depth of knowledge about the impacts of climate change on public health. This workshop was designed to identify priorities across the environmental and health sectors and begin the discussion on building a public health sector response to address the impacts of climate change on drinking water and human health.

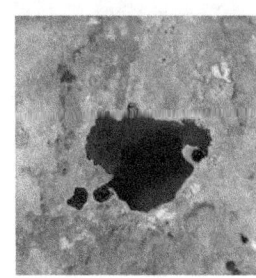

Effects of Climate Change on Human Health—North Carolina. This study linked 10 years of data related to daily asthma and myocardial infarction hospital admissions, air quality, and weather patterns in a number of cities in North Carolina (see <cfpub.epa.gov/ncer_abstracts/index.cfm/fuseaction/display.abstractDetail/abstract/7887/report/2006>). The effect of eight air mass types, classified using the Spatial Synoptic Classification system, on ozone and particulate matter (PM) concentrations and subsequent processes was evaluated. Preliminary results show that the distributions of air pollutant concentrations are different under different air masses. In addition, the analysis found that ozone and PM10 are positively related to asthma admissions for some cities under the Dry Tropical and the Dry Moderate air masses.

Valley Fever Public Health Decision-Support System. Valley Fever (coccidioidomycosis) is a disease endemic to arid regions in the Western Hemisphere, and is caused by soil-dwelling fungi. Arizona is currently experiencing an epidemic, with almost 4,000 cases in 2004, greatly exceeding other climate-related diseases such as hantavirus pulmonary syndrome or West Nile fever virus in the United States. The fungus responds to changes in climate conditions, such as precipitation and atmospheric dust. Climate models and satellite-derived spatial data on soil moisture and land-cover disturbance are being used to evaluate seasonal associations with Valley Fever incidence. Working in partnership with the Arizona Department of Health Services, the University of Arizona is developing a decision-support tool to make seasonal forecasts of disease incidence by geographic area and display spatial relationships with environmental conditions (see <cfpub.epa.gov/ncer_abstracts/index.cfm/fuseaction/display.abstractDetail/abstract/7885/report/00>).

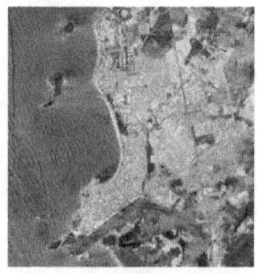

HIGHLIGHTS OF HUMAN CONTRIBUTIONS AND RESPONSES TO ENVIRONMENTAL CHANGE – PLANS FOR FY 2009

Climate Change and Human Health. Exposure to allergens results in allergenic illnesses in approximately 20% of the U.S. population. Climate change, including increased atmospheric carbon dioxide (CO_2) concentrations, could have significant impacts on the production, distribution, dispersion, and allergenicity of aeroallergens and the growth and distribution of organisms that produce them (i.e., weeds, grasses, trees, and fungus). Shifts in aeroallergen production and, subsequently, human exposures, may result in changes in the prevalence and severity of symptoms in individuals with allergenic illnesses. EPA and CDC plan to investigate this potential health effect through the award of competitive multi-year grants as part of an interagency program on climate change and health.

This activity will address Question 9.4 of the CCSP Strategic Plan.

Climate Change and Water Quality. The movement of water through the atmosphere, the exchange of water between the atmosphere and the surface, and the movement and storage of water on and below the land surface are linked through physical and dynamical processes occurring over a variety of spatial and temporal scales. Watershed biogeochemical and other processes, including interaction with human stressors such as changes in land use, pollutant loading, and water flow management, likewise occur over a variety of spatial and temporal scales. Climate change has the potential to interact with these systems and stressors in complex ways, leading to potentially significant impacts on water quality. Cutting-edge research, conducted through a competitive grants program, will be undertaken to improve information and understanding regarding the ways climate change affects water quantity and quality that in turn will lead to better analysis and decisionmaking.

This activity will address Question 9.1 of the CCSP Strategic Plan.

Exploring the Role of Science and Technology in Climate Change Mitigation and Adaptation. A significant challenge for models that simulate human dimensions of climate change is their ability to incorporate the potential role of scientific advances into the mix and what this might mean for transformational shifts in technology, both to mitigate and adapt to climate change. Similarly, improved probabilistic frameworks and alternative methods are needed that provide complementary perspectives within and outside of existing integrated assessment models. During 2007, a workshop attended by many agencies and international researchers focused attention on this vexing challenge and offered alternative perspectives on methods and tools that should be considered. Two national laboratory teams and one university-based team have been challenged to take on this assignment at, recognizably, a modest level of effort. Recognizing the highly important role and corresponding uncertainties of potential advances in science and technology, especially as informed by SAP 2.1, additional analytical, computational, and mathematical expertise will be pursued in subsequent years as funding permits.

This activity will address Question 9.1 of the CCSP Strategic Plan.

Research into the Energy/Water Nexus. Climate change projections suggest that water availability is likely to be affected in coming decades, especially in regions whose surface water supply depends substantially on winter snowfall. The national laboratories, in partnership with universities and industry, are exploring new techniques for incorporating water demand and management strategies within the framework of integrated assessment models, particularly to understand the sensitivities of biomass production, agricultural land use, and energy production to adaptation and management decisions. Investigations are attempting to unravel a host of connections between such potential regional water-supply impacts of climate change and the Nation's energy systems. Examples range from effects on energy supply as climate change increases demands for cooling—related to such issues as hydropower resources, cooling water requirements for thermal electricity generation, and water requirements for bioenergy production—to energy needs for groundwater pumping and surface water transport, all of this in a context of demographic, economic, and land-use change. Additional future research will attempt to link improved precipitation modeling capacities in Earth system models with integrated assessment models and analyses of climate change impacts and adaptation potentials in order to anticipate needs for decision support related to possible pressures on the "energy/water nexus."

This activity will address Question 9.2 of the CCSP Strategic Plan.

Understanding Infrastructure and Energy Vulnerabilities under Extreme Weather Challenges. In many cases, social concerns about the effects of climate change are focused on extremes rather than averages. Examples include severe storms, heat waves, and droughts. In addition, there is significant concern that many important ecosystem responses—such as the response of forests and crops to pests, pathogens, and fire—can be quite abrupt, as seen now in the pine bark beetle epidemics in the Pacific Northwest and the increase in fire frequency throughout the West. The integrated assessment community is developing methods for incorporating our best current understanding of such phenomena in the context of adaptation decisions. Close interactions with Earth system modelers and the ecological research community are important components of this research.

In the case of energy systems, particular concerns include effects of heat waves on electricity demand and distribution systems (as in the summer of 2006), effects of droughts on competition among economic sectors for scarce water, and effects of severe storms on energy infrastructures in vulnerable areas (as with Hurricane Katrina). One area of active research in the Earth system modeling community is enhancing the ability of climate models to project extremes. Linked to these developments and with regard to the human dimensions, research efforts are beginning that will improve capacities to understand and model implications of such extreme events for energy systems and infrastructures and their possible effects on national and regional economies.

This activity will address Question 9.2 of the CCSP Strategic Plan.

Highlights of Recent Research
in Decision-Support Development

One of the main purposes of CCSP is to provide information for decisionmaking through the development of decision-support resources. Decision-support resources include analyses and assessments, interdisciplinary research, analytical methods (including scenarios and alternative analysis methodologies), model and data product development, communication, and operational services that provide timely and useful information to address questions confronting policymakers, resource managers, and other stakeholders. This research is especially relevant to CCSP Goal 5: "Explore the uses and identify the limits of evolving knowledge to manage risks and opportunities related to climate variability and change."

Decision-support resources are targeted at three broad categories of uses: (1) discussion and planning based on state-of-the-science syntheses and assessments by decisionmakers, stakeholders, the media, and the general public; (2) operational adaptive management decisions undertaken by managers of natural resources and built infrastructure (i.e., "climate services applications"); and (3) climate change policy formulation. Each of these categories has a unique set of stakeholders and requires different decision-support tools. However, they share a common reliance on partnerships between scientists and stakeholders to define the problems to be addressed, the nature of decision-support resources to be developed, the expected information to be provided, and the approach for describing levels of confidence and key uncertainties.

Development of decision-support resources cannot be isolated in a single program element, disconnected from research throughout CCSP. Responsibility for developing decision-support resources is distributed across CCSP and success depends on developing strategies for integrating knowledge from the many diverse fields represented in the program. At the same time, CCSP's strategy for improving understanding of human-environment interactions recognizes the need for basic research into both the natural sciences and the human dimensions of global change that may not lead directly to decision-support resources.

A primary activity to meet CCSP Decision Support Goal 1 (see page 122) is development of 21 synthesis and assessment products to support informed decisionmaking on climate variability and change by a broad group of stakeholders, including policymakers, resource managers, media, and the general public. The development of these SAPs stems from the Global Change Research Act (GCRA) of 1990 (P.L. 101-606, section 106), which directs the program to "...produce information readily usable by policymakers attempting to formulate effective strategies for preventing, mitigating, and adapting to the effects of global change" and to undertake periodic science "assessments." A complete list of the 21 SAPs, by title, is given on page 29. Status updates can be obtained from <www.climatescience.gov/Library/sap/sap-summary.php>, including information on opportunities for public comment on draft products. Descriptions of those SAPs most relevant to either Human Contributions and Responses or Decision-Support Resources Development are described in the accompanying box.

The National Research Council's Committee on the Human Dimensions of Global Change. An important source of scientific expertise and judgment on societal issues related to global change is the NRC Committee on the Human Dimensions of Global Change (CHDGC). The committee was formed in 1989 to help guide U.S. research on the

SYNTHESIS AND ASSESSMENT PRODUCTS

Human Contributions and Responses

SYNTHESIS AND ASSESSMENT PRODUCT 2.1
Scenarios of Greenhouse Gas Emissions and Atmospheric Concentrations (Part A) and Global-Change Scenarios: Their Development and Use (Part B)
This product, which was publicly released in October 2007, provides a new long-term, global reference for greenhouse gas stabilization scenarios and an evaluation of the process by which scenarios are developed and used. SAP 2.1 consists of two parts. Part A uses computer-based scenarios to evaluate four alternative stabilization levels of greenhouse gases in the atmosphere and the implications for energy and the economy of achieving each level. Part A includes stabilization scenarios for the six primary anthropogenic greenhouse gases—carbon dioxide, nitrous oxide, methane, hydrofluorocarbons, perfluorocarbons, and sulfur hexafluoride—and it uses updated economic and technological data and new tools for scenario development. Although these scenarios should not be considered definitive predictions of future events, they provide valuable insights for decisionmakers. Part B examines how scenarios have been developed and used in global climate change applications, evaluates the effectiveness of current scenarios, and recommends ways to make future scenarios more useful. Part B of the report concludes that scenarios can support decisionmaking by providing insights regarding key uncertainties, including future emissions and climate as well as other environmental and economic conditions.

SYNTHESIS AND ASSESSMENT PRODUCT 4.1
Coastal Elevations and Sensitivity to Sea-Level Rise
This product will examine the vulnerability of coastal areas in the U.S. mid-Atlantic states to sea-level change. Specific questions to be addressed include identifying which areas are low enough to be inundated by tides, how floodplains would change due to a changing climate, which areas might be subject to erosion, and locations where wetlands will be able to migrate inland versus locations where shores will be protected. The product will examine the implications of sea-level rise, including impacts on population and economic activity in vulnerable areas, costs of shore protection, ecological effects, flood damages, public access to modified shore areas, cases where sea-level rise justifies policy changes, options being considered by conservancies and governments, and lessons from the unfolding consequences of the 2005 hurricanes in the Gulf Coast region.

SYNTHESIS AND ASSESSMENT PRODUCT 4.2
Thresholds of Change in Ecosystems
There is a body of ecosystems research that focuses on enhancing understanding of climate change impacts on ecosystems (and *vice versa*) and developing the capability to predict potential impacts of future climate change. Increasing emphasis is being placed on climate-related thresholds that could result in discontinuities or sudden changes in ecosystems and climate-sensitive resources. Discontinuities in responses of ecosystems and resources are difficult to predict, and may significantly affect human societies that depend on ecosystem goods and services. Improved understanding of such sudden changes is essential to managing ecosystems and resources in the face of climate change. This report will synthesize the present state of scientific understanding regarding thresholds of change that trigger sudden changes in ecosystems and climate-sensitive resources. The report will develop a conceptual framework for characterizing sudden changes, and synthesize peer-reviewed studies that provide the best available evidence for defining circumstances that trigger discontinuities in response to climate change.

interactions between human activity and global environmental change. CHDGC focuses on two main tasks: developing the intellectual basis for progress in understanding human-environment interactions, and advising on future research directions. CHDGC recently completed a study entitled *Research and Networks for Decision Support in the NOAA Sectoral Application Research Program*. The Sectoral Applications Research Program (SARP) focuses on the needs for climate-related information to inform decisions in particular "sectors," defined by resources (e.g., coastal, water resources, forests, agricultural lands) or by decision domains (e.g., emergency management, urban planning). SARP is a new program and CHDGC was asked to provide advice on the role that SARP could play, the best approaches to meet program goals, and how to monitor and evaluate the effectiveness of SARP. The report, which was released in

SYNTHESIS AND ASSESSMENT PRODUCTS (CONTINUED)

Human Contributions and Responses (continued)

SYNTHESIS AND ASSESSMENT PRODUCT 4.3
The Effects of Climate Change on Agriculture, Land Resources, Water Resources, and Biodiversity

This report, publicly released in May 2008, addresses the effects of climate change on agriculture, forestry, land and water resources, and biodiversity. Air and water temperature, precipitation, and related climate variables are fundamental regulators of biological processes. For this reason, human-induced climate change has the potential to affect the condition, composition, structure, and function of ecosystems. Such changes may also alter the linkages and feedbacks between ecosystems and the climate system. Additionally, ecosystems produce a wide array of goods and services valued by humans and in many cases essential for human survival and property. Climate-related changes in ecosystems and other key resources could have impacts on human communities and economic conditions.

SYNTHESIS AND ASSESSMENT PRODUCT 4.4
Preliminary Review of Adaptation Options for Climate-Sensitive Ecosystems and Resources

Climate is a dominant factor influencing the distribution, abundance, structure, and function of, and services provided by, ecosystems. Many ecosystems are thus vulnerable to future changes in climate. The goal of adaptation is to reduce these risks of adverse ecological outcomes through management activities that increase the resilience of these systems to climate change. Resilience is defined here as the magnitude of disturbance that can be absorbed by a system before it shifts from one stable state (or stability domain) to another and the speed of return of a system to equilibrium after a disturbance has occurred. This report, which was publicly released in June 2008, provides a review and synthesis of information on adaptation options for selected climate-sensitive ecosystems in order to aid in designing management strategies that facilitate adaptation, provide examples of how to implement strategies in specific places, and identify issues and challenges associated with implementation of adaptation options.

SYNTHESIS AND ASSESSMENT PRODUCT 4.5
Effects of Climate Change on Energy Production and Use in the United States

This report, which was publicly released in October 2007, summarizes current knowledge of the potential effects of climatic change on energy production and use in the United States. It focuses on three questions: (1) How might climatic change affect energy use in the United States, (2) how might climatic change affect energy production and supply in the United States, and (3) how might climatic change have other effects that indirectly shape energy production and use in the United States? Great care was taken in answering these questions, for two reasons. One, the available research literature on these key questions is limited, supporting a discussion of issues but not providing definite answers. Two, as with many other aspects of potential effects of climatic change on the United States, the effects on energy production and use depend on more than climatic change alone; other potentially important factors include patterns of economic growth and land use, patterns of population growth and distribution, technological change, and social and cultural trends that could shape policies and actions, individually and institutionally.

SYNTHESIS AND ASSESSMENT PRODUCTS (CONTINUED)

Human Contributions and Responses (continued)

SYNTHESIS AND ASSESSMENT PRODUCT 4.6

Analyses of the Effects of Global Change on Human Health and Welfare and Human Systems

This product will examine the effects of global change on human systems. It will address Goal 4 of the *CCSP Strategic Plan*: to "understand the sensitivity and adaptability of different natural and managed ecosystems and human systems to climate and related global changes". The impacts of climate variability, climate change, shifting patterns of land use, and changes in population patterns are human problems, not simply problems for the natural or the physical world. This SAP will examine the vulnerability of human health and socioeconomic systems to global environmental change across three areas of potential impacts and adaptations: human health, human settlements, and human welfare. It will address the questions of what, where, and when climate variability and change will affect U.S. social systems. The challenge for this project will be to assess risks associated with health, welfare, and settlements and to identify and develop timely adaptive strategies to address human vulnerabilities. The primary goals for adaptation to climate change and variability focus on managing significant risks proactively when possible; establishing protocols to detect and measure risks; and leveraging technical and institutional adaptive capacity to address new climate risks, especially as they exceed conventional adaptive measures.

SYNTHESIS AND ASSESSMENT PRODUCT 4.7

Impacts of Climate Variability and Change on Transportation Systems and Infrastructure: Gulf Coast Study

This report, which was publicly released in March 2008, addresses the potential effects of climate variability and change on transportation infrastructure and systems in the central Gulf Coast of the United States. The purpose of this study was to increase the knowledge base regarding the risks and sensitivities of transportation infrastructure to climate variability and change, the significance of these risks, and the range of adaptation strategies that may be considered to ensure a robust and reliable transportation network. Implications for all transportation modes—surface, marine, and aviation—are addressed. This SAP is a case study that focuses on the Gulf Coast, and assesses the significant risks to transportation, develops methodology to be applied in other geographic locations, identifies potential strategies for adaptation, and develops decision-support tools to assist transportation decisionmakers in incorporating climate-related trend information into transportation system planning, design, engineering, and operational decisions.

Decision-Support Resources Development

SYNTHESIS AND ASSESSMENT PRODUCT 5.1

Uses and Limitations of Observations, Data, Forecasts, and Other Projections in Decision Support for Selected Sectors and Regions

This product will focus on characterizing a subset of the observations from remote-sensing and *in situ* instrumentation that are of high value for decisionmaking. The product will characterize observational capabilities that are currently or could potentially be used in decision-support tools, catalog a subset of ongoing decision-support activities that use these capabilities, and evaluate a limited number of case studies of these decision-support activities. The detailed evaluation of decision-support activities and demonstration projects will provide information to agencies and organizations responsible for developing, operating, and maintaining selected decision-support processes and tools. The evaluation will also provide information on the nature of interactions between users and producers of climate science information, approaches for accessing science information, and assimilation of scientific information in the decisionmaking process. The product will include an online catalog of decision-support demonstration projects with interactive links, which will be updated as additional experiments are conducted and new approaches to incorporating and benchmarking application of observations and other global change research products evolve.

SYNTHESIS AND ASSESSMENT PRODUCT 5.2

Best Practice Approaches for Characterizing, Communicating, and Incorporating Scientific Uncertainty in Decisionmaking

This product will address the issue of uncertainty and its relationship to science, assessment, and decisionmaking. Specifically, the product is intended to help improve the quality and consistency of information about scientific uncertainty presented to decisionmakers and other users of CCSP reports by identifying "best practice" options recommended in the literature on this subject; to improve communication between scientists and users of the products by providing recommendations for addressing uncertainty; and to provide a brief overview of the literature on approaches for communicating and considering uncertainty related to climate.

SYNTHESIS AND ASSESSMENT PRODUCTS (CONTINUED)

Decision-Support Resources Development (continued)

SYNTHESIS AND ASSESSMENT PRODUCT 5.3

Decision Support Experiments and Evaluations using Seasonal to Interannual Forecasts and Observational Data

This product will concentrate on the water-resource management sector. It will describe and evaluate current forecasts, assess how forecasts are being used in decision settings, and evaluate decisionmakers' level of confidence in these forecasts. The participants in the development of this product (primarily consisting of government officials, researchers, and users) will evaluate forecasts as well as their delivery in order to identify options for improving partnerships between the research and user communities. It will inform decisionmakers about the experiences of others who have experimented with the use of seasonal and interannual forecasts and other observational data; climatologists and social scientists about how to advance the delivery of decision-support resources that use the most recent forecast products, methodologies, and tools; and science managers as they plan for future investments in research related to forecasts and their role in decision support.

September 2007, recommended specific research, workshop, and pilot project activities as ways to carry out the SARP program. The NRC study further recommended principles for selecting activities within each sector, including promotion of social innovation in using climate science, high-impact decisions, leveraging investments through partnerships, fertile ground, increasing resilience and adaptability, equity, and research of interest to social science.

SAP 2.1a—Scenarios of Greenhouse Gas Emissions and Atmospheric Concentrations. SAP 2.1a was produced under CCSP auspices by an independent advisory committee to one of the participating agencies under the provisions of the Federal Advisory Committee Act (FACA). The report applies three computer-based integrated assessment models in a comparison of five different scenarios of greenhouse gas emissions under alternative assumptions regarding long-term, global climate goals. The CCSP report is the first to use several alternative models to evaluate multiple stabilization scenarios in this way. SAP 2.1a produced numerous findings. One of the most important implications of the work reported was the primacy of technology in addressing climate change, not only in the near term, but also in the long term where investments in basic science and technology can lay the foundations for dramatically improved technologies to deploy.

Decision Assessment in the Gulf Coast and Chesapeake Bay Regions. Pilot studies in the Gulf Coast region and the Chesapeake Bay were undertaken to test different approaches to assessing the flow and use of climate change science information in decisionmaking, the factors and institutions that affect its use, and the types and characteristics of decisions most sensitive to climate change and most in need of additional reevaluation and research in light of projected changes. Results from these studies are being used to determine the applicability of a decision assessment approach to the national level and to decisions related to water quality.

133

Integrated Evaluation of Climate Change, Mitigation, Bioenergy, and Land Use. The MIT Joint Program on the Science and Policy of Global Change completed its linkage of a multi-sector, multi-region general equilibrium model of the world economy with a terrestrial ecosystem model that simulates biogeochemical processes of land systems at a 0.5° latitude-longitude grid level. Additionally, MIT conducted initial testing and sensitivity analyses using these enhanced capabilities focusing on the effects of disturbances associated with the conversion of unmanaged forest and grassland to crop, pasture, and bioenergy production on greenhouse gas concentrations. Preliminary analysis confirms that second-generation biofuels technology could be an important mitigation option. However, depending on the nature of the land supply response and for some scenarios, biofuels also could lead to substantial deforestation and release of carbon. Looking toward the future, these new science-based modeling capabilities, and subsequent refinements, may prove valuable tools in helping to inform strategies to increase the benefits from biofuels while attempting to understand and manage to reduce unintended, undesirable consequences. The modeling facility will be further enhanced in 2009 to examine the complex interactions of Earth system change, agriculture and energy markets, and mitigation and adaptation.

Improving the Methodological Science Base of Integrated Assessment Research. The integrated assessment community sponsored a workshop in the summer of 2007 focused primarily on new methods for coupling integrated assessment models with Earth system models, and understanding the needs for mathematical and computational advances to achieve such. Topics of particular importance were the need for techniques for quantitative characterization of uncertainty in both model parameterization and model structure, the need to characterize the actual decisions that the models are meant to inform, and understanding of the interdisciplinary context within which decisions are to be made.

Development of Modeling Tools to Support Water and Watershed Management. Climate change presents a range of risks and opportunities to water managers. Managing these risks requires an improved understanding of potential impacts, and strategies for increasing the resilience of water and watershed systems to anticipated change. Given the uncertainty in projections of future climate at local and regional scales, water managers need to develop management strategies robust to the full range of plausible conditions and events. To enable managers to develop such strategies, a new climate assessment capability was developed within the Better Assessment Science Integrating Point and Non-point Sources (BASINS) watershed

modeling system. To demonstrate the tools' capabilities, a case study using the new BASINS system was undertaken to assess the sensitivity of hydrologic and water quality endpoints to climate change. The hydrologic and water quality endpoints used in the climate sensitivity study included mean annual streamflow; the 100-year flood event; 7Q10 low stream flow (lowest consecutive 7-day streamflow likely to occur in a 10-year period); and mean annual sediment, phosphorus, and nitrogen loads. The climate sensitivity study was conducted in the Monocacy River watershed.

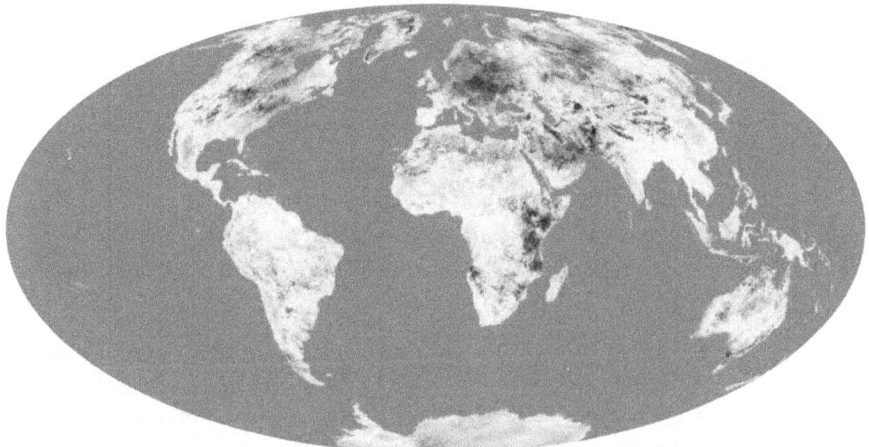

Interagency Workshop on Decision Support for Heat Wave Preparedness. **Excessive** heat events are, and will continue to be, a public health threat in the United States. Empirical data demonstrate that these events often increase the number of daily deaths (mortality) and other nonfatal adverse health outcomes (morbidity) in affected populations. The number of extremely hot days is expected to increase; however, there are a number of low cost but effective responses that could reduce the health impacts of these events. An interagency workshop, *Excessive Heat Events: Confronting Climate Change, Vulnerability, and Urbanization by Improving Heat Health Services, Mitigation Strategies, and Communications*, was convened in November 2007. The goal of the workshop was to engage stakeholders—Federal, State, and local government agencies, academia, industry, and professional interests—in an effort to identify public health priorities, data gaps, and decision-support tools to enhance local responses to heat waves.

Adaptation Guidebook. The Climate Impacts Group (CIG), a CCSP-supported Regional Integrated Sciences and Assessments (RISA) team, and King County, Washington's Climate Team created a guidebook on preparing for and adapting to climate change. *Preparing for Climate Change: A Guidebook for Local, Regional, and State Governments* is designed to help local, regional, and State governments prepare for climate

change by recommending a detailed, easy-to-understand process for climate change preparedness based on familiar resources and tools. The International Council for Local Environmental Initiatives (ICLEI) Local Governments for Sustainability contributed to the production and dissemination of the guidebook to make it accessible to local governments across the United States. The results of this guidebook will be used in future planning for CCSP-supported sectoral work on urban issues.

Decision-Support Workshop for Coastal Extension. CCSP convened a decision-support *Workshop on Climate Science and Services: Coastal Applications for Decision Making through Sea Grant Extension and Outreach* in Charleston, South Carolina (April 2007). Scientists associated with and supported by CCSP agencies shared important insights about impacts and adaptation, and potential methods for using climate information in decisionmaking. The workshop marks an important step in the development of an expanded partnership among CCSP's climate and coastal programs in an effort to provide enhanced support and services for national, State, and local constituencies concerned with coastal resource management and planning in the face of a dynamic climate system. Workshop presentations and related materials can be found at <csc.noaa.gov/sgcw>.

Decision-Support Research on Water Resources. CCSP continues to advance research on the linkages between climate and key sectors, including water resources. A sampling of the activities nearing or achieving completion include (1) a project by Stratus Consulting that is concentrating on the potential effects of climate change in combination with a repeat of long-term climate variability in Boulder, Colorado, and (2) a project recently completed by the National Drought Mitigation Center (NDMC) involving work with farmers to better understand the linkages between sustainable agriculture and drought management. Among the topics analyzed were the agricultural practices implemented to reduce the effects of drought, how drought information has been incorporated into agricultural management, and how drought and climate products could be enhanced to meet the needs of producers. The results of this study will be used both within the NDMC as well as for National Integrated Drought Information System (NIDIS) planning.

HIGHLIGHTS OF DECISION-SUPPORT DEVELOPMENT — PLANS FOR FY 2009

Upcoming Report from the Committee on the Human Dimensions of Global Change. CHDGC is expected to release *Strategies and Methods for Climate-Related Decision Support* in FY 2009. This report will elaborate a framework for organizing and evaluating decision-support activities for CCSP, with special attention to sectors and issues of concern to the sponsors. It will also consider needs for science in support of decisions related to natural disasters and extreme events associated with climate change, such as droughts, floods, and hurricanes. The study panel will consider the range of relevant decisions, decisionmakers, decision contexts, spatial and temporal frames, and decision-support objectives, and current and potential strategies for organizing decision-support efforts to meet these objectives—taking into account the fact that, in some sectors, the desired outcomes of decision-support activities may not be clear in advance.

This activity will support Decision Support Goals 1 and 3 of the CCSP Strategic Plan.

Testing, Inter-Model Comparison, and Validation Methods for Integrated Assessment Research. There is a recognized need to improve testing, inter-model comparisons, and validation of integrated assessment models as the rapidly expanding field assumes greater prominence and importance in helping to inform, at the national and regional levels, climate change policies and actions over the coming years. SAP 2.1 was a first of its kind U.S. effort that involved the comparison of three models from three independent teams: MIT, JGCRI, and the Electric Power Research Institute (EPRI). It marked an important beginning that, through subsequent discussions, has highlighted the many advanced techniques and methods that can and will be applied to help strengthen the scientific discipline and rigor within the field. The topic took on greater focus at the 2007 Integrated Assessment Annual Meeting with participation of many agencies and domestic and international researchers. The need for testing and validation was also separately highlighted by an independent advisory committee to one of the lead agencies that sponsors integrated assessment research. Several potential paths forward were discussed at the annual meeting and planning is underway to pursue select targets of opportunity. Close coordination with the Earth system modeling community is anticipated as this initiative takes shape.

This activity will support Decision Support Goal 3 of the CCSP Strategic Plan.

Modeling Tool to Enable Assessment of Soil Erosion. An online decision-support capability within the USDA Agricultural Research Service's Water Erosion Prediction Project (WEPP) soil erosion model is being developed. New climate change assessment capabilities within WEPP will enable land managers to develop best management practices to lessen the impacts of climate variability and change on sediment loading

from agricultural land to streams. The need for developing similar climate assessment capabilities for models applicable to urban drainage and design will be evaluated.

This activity will support Decision Support Goal 2 of the CCSP Strategic Plan.

Coping with Drought. CCSP, working with its partner Federal agencies through regional and sectoral projects, will support research and stakeholder interactions focused on using climate impacts information for drought planning and resource management. Emphasis will be placed on drought-prone areas of the western and southeastern United States where conflicts over water are growing. This will contribute to both the NIDIS being developed across a number of CCSP agencies and to the CCSP FY 2009 priority on climate, hydrology, and water management.

This activity will support Decision Support Goal 2 of the CCSP Strategic Plan.

Disaster Preparedness. The CCSP-sponsored International Research Institute for Climate and Society has entered into a partnership with the International Federation of Red Cross and Red Crescent Societies to use climate information and forecasts to improve disaster preparedness and response capabilities. Traditional climate tools are being tailored in innovative ways to identify areas that are especially vulnerable to climate-related disasters.

This activity will support Decision Support Goal 2 of the CCSP Strategic Plan.

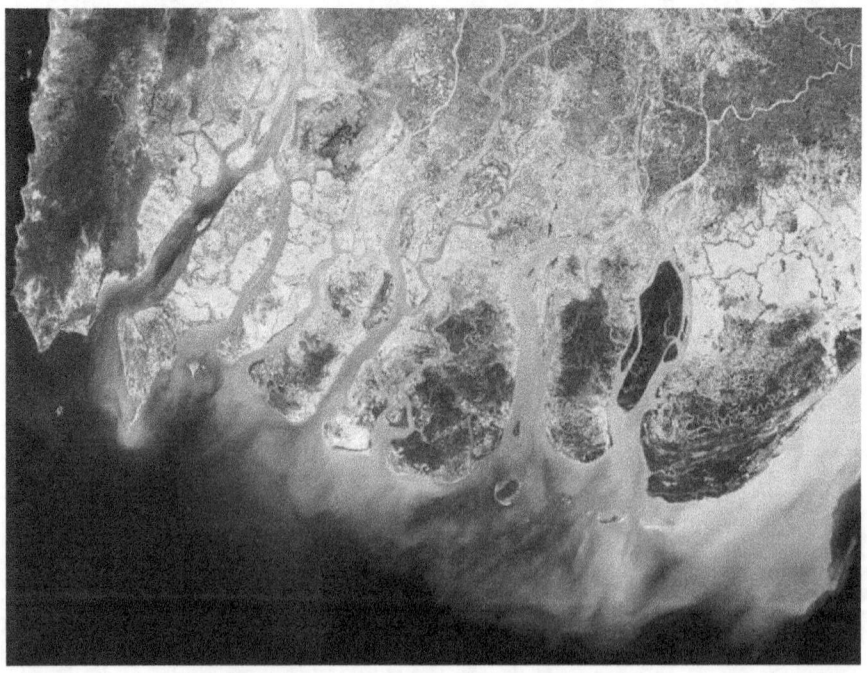

Decision Support for Coastal Resource Management and Community Resilience. As part of a broader effort, CCSP plans to develop decision-support resources and research applications for coastal resource management and hazards preparedness through place-based interdisciplinary climate research and assessment focused on vulnerability, impacts, and adaptation. Examples of decision-support resources include web portals; derived data analysis products; geographic information system (GIS) tools that integrate social, economic, and climate data in a useful and interactive format; "guidebooks" for public distribution describing potential climate sensitivity in coastal regions; impacts assessments; and methodologies for applying climate information in management and policy activities related to coastal resources.

This activity will support Decision Support Goal 2 of the CCSP Strategic Plan.

HUMAN CONTRIBUTIONS AND RESPONSES AND DECISION-SUPPORT RESOURCES DEVELOPMENT CHAPTER REFERENCES

1) **NRC**, 2001: *Climate Change Science: An Analysis of Some Key Questions.* National Academy Press, Washington, DC, USA, 29 pp.

2) **IPCC**, 2007: Summary for Policymakers. In: *Climate Change 2007: The Physical Science Basis. Contribution of Working Group I to the Fourth Assessment Report of the Intergovernmental Panel on Climate Change* [Solomon, S., D. Qin, M. Manning, Z. Chen, M. Marquis, K.B. Averyt, M. Tignor, and H.L. Miller (eds.)]. Cambridge University Press, Cambridge, United Kingdom and New York, NY, USA.

3) **Snover**, A.K., L. Whitely Binder, J. Lopez, E. Willmott, J. Kay, D. Howell, and J. Simmonds, 2007: *Preparing for Climate Change: A Guidebook for Local, Regional, and State Governments.* In association with and published by ICLEI – Local Governments for Sustainability, Oakland, CA. Available at <cses.washington.edu/cig/fpt/guidebook.shtml>.

8 | Observing and Monitoring the Climate System

Observing and Monitoring the Climate System

Goal 12.1: Design, develop, deploy, and integrate observation components into a comprehensive system.
Goal 12.2: Accelerate the development and deployment of observing and monitoring elements needed for decision support.
Goal 12.3: Provide stewardship of the observing system.
Goal 12.4: Integrate modeling activities with the observing system.
Goal 12.5: Foster international cooperation to develop a complete global observing system.
Goal 12.6: Manage the observing system with an effective interagency structure.

Data Management and Information

Goal 13.1: Collect and manage data in multiple locations.
Goal 13.2: Enable users to discover and access data and information via the Internet.
Goal 13.3: Develop integrated information data products for scientists and decisionmakers.
Goal 13.4: Preserve data and information.

See Chapters 12 and 13 of the *Strategic Plan for the U.S. Climate Change Science Program* for detailed discussion of these goals.

Two overarching questions are identified in the CCSP Strategic Plan for "Observing and Monitoring the Climate System" and "Data Management and Information". These questions continue to offer guidance to these elements of the program:

- How can we provide active stewardship for an observation system that will document the evolving state of the climate system, allow for improved understanding of its changes, and contribute to improved predictive capability for society?
- How can we provide seamless, platform-independent, timely, and open access to integrated data, products, information, and tools with sufficient accuracy and precision to address climate and associated global changes?

High-quality, long-term observations of the global environment are essential for defining the current state of the Earth's environmental system, its history, and its variability. This task requires both space- and surface-based observation systems. Climate observations encompass a broad range of environmental observations, including (1) routine weather observations, which are collected consistently over a long period of time; (2) observations collected as part of research investigations to elucidate processes that contribute to maintaining climate patterns or their variability; (3) highly precise, continuous observations of climate system variables collected for the express purpose of documenting long-term (decadal to centennial) change; and (4) observations of climate proxies, collected to extend the instrumental climate record to remote regions and back in time.

The United States contributes to the development and operation of several global observing systems, both research and operational, that collectively provide a comprehensive measure of climate system variability and climate change processes. These systems are a baseline Earth-observing system and include NASA, NOAA, and USGS Earth-observing satellites and extensive *in situ* observational capabilities. CCSP also supports several ground-based measurement activities that provide the data used in studies of the various climate processes necessary for better understanding of climate change. U.S. observational and monitoring activities contribute significantly to several international observing systems, including the Global Climate Observing System (GCOS) principally sponsored by the World Meteorological Organization (WMO); the Global Ocean Observing System sponsored by the United Nations Educational, Scientific, and Cultural Organization's Intergovernmental Oceanographic Commission (IOC); and the Global Terrestrial Observing System sponsored by the United Nations Food and Agriculture Organization. The latter two have climate-related elements being developed jointly with GCOS.

A specific subset of the GCOS observing activities for 2007 and 2008 (and into 2009) are the CCSP-sponsored polar climate observations made in cooperation with the International Polar Year (IPY). During 2009, IPY will come to a formal conclusion; however, many polar observing systems will continue to operate. Several agencies are working together to establish an Arctic Observing Network that will build on systems deployed during IPY and provide for coordinated efforts to sustain key climate observations. This cooperation will extend to international partners to encourage a pan-Arctic approach to observation and data sharing.

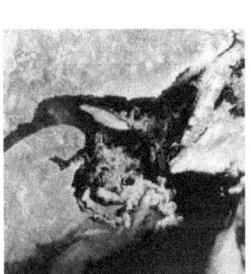

Remotely sensed observations continue to be a cornerstone of CCSP. The Cloud-Aerosol Lidar and Infrared Pathfinder Satellite Observations (CALIPSO) lidar and CloudSat radar instruments are providing an unprecedented examination of the vertical structure of aerosols and clouds over the entire Earth. These data—when combined with data from the Aqua, Aura, and Parasol satellites orbiting in formation (the "A-Train")—will enable systematic pursuit of key issues including the effects of aerosols on clouds and precipitation, the strength of cloud feedbacks, and the characteristics of difficult-to-observe polar clouds. The increasing volume of data from remote-sensing and *in situ* observing systems presents a continuing challenge for CCSP agencies to ensure that data management systems are able to handle the expected increases.

HIGHLIGHTS OF RECENT RESEARCH— OBSERVATIONS AND MONITORING

The following are selected highlights of observation and monitoring activities supported by CCSP-participating agencies. The principal focus of this chapter is on describing progress in implementing the observations that contribute to the CCSP mission. As a result, the chapter touches on some observing systems that are crucial to CCSP but are not included within the CCSP budget because they primarily serve other purposes.

Integrated Surface Climate Observations. The integration of a series of surface observing system networks is intended to sustain the Nation's record of land surface measurements essential to monitor and assess the surface climate. This project integrates land surface observations from regional, national, and international sources. Three major surface climate networks cover the U.S. region: (1) the U.S. Climate Reference Network (CRN) sites; (2) the Surface Energy Budget Network; and (3) a modernized Historical Climate Network. CCSP's automated observing systems, voluntary cooperative observing systems, mesonet observing systems, and private sector observations are also integrated into the system. These surface observing systems contribute measurements of 10 key GCOS essential climate variables: air temperature, precipitation, atmospheric pressure, surface radiation, vector winds, water vapor, clouds, soil temperature, soil moisture, and snow depth.

Initial Ocean Observing System for Climate Reaches 59% Completion. CCSP agencies cooperate with 66 nations in implementing the internationally vetted design of an initial ocean observing system for climate, articulated in the WMO/IOC/UNEP plan for GCOS. Deployment of the observing system, planned for completion in 2013, is proceeding, with the United States currently supporting nearly 50% of the ocean-based observing platforms.

Tropical Moored Buoy Network Extended into the Indian Ocean. CCSP continues to provide leadership in the development of the Indian Ocean Observing System (IndOOS), a multi-national, multi-platform network designed to support climate forecasting and

research. IndOOS is a regional cornerstone of the Global Earth Observing System of Systems (GEOSS) and has been endorsed by committees of the World Climate Research Programme and the IOC. By the end of FY 2008, the array of moorings will consist of 12 sites (four new sites in FY 2008), bringing the total moored buoy array, including all international contributions, to 43% completion.

GCOS Tide Gauge Network Expands Real-Time Reporting Capacity. The GCOS tide gauge network is a subset of the global sea-level observing system, providing high-precision, geo-located tide gauge records appropriate for monitoring long-term climate trends in sea level. Plans call for a network of 180 stations across the globe. As of the beginning of FY 2008, the United States and its partners had added, upgraded, or maintained 127 stations in support of the international goal. The reference-level data sets are used in conjunction with operational numerical models for the calibration of satellite altimeter data, the compilation of oceanographic data products, and research on interannual to decadal climate fluctuations and short-term extreme events. They are also used by various national tsunami warning agencies for tsunami monitoring.

Global Coverage Achieved by the Argo Profiling Array. In 1998, an international consortium presented plans for an array of 3,000 autonomous instruments that would revolutionize the collection of climate-relevant information from the upper 2 km of the world's oceans—the Argo array. These instruments drift at depth, periodically rising to the sea surface, collecting data along the way, and report their observations in real-time via satellite communications. The initial deployment objective of 3,000 instruments distributed homogenously throughout the world's oceans has been attained and the array now provides over 100,000 high-quality temperature and salinity profiles annually along with global-scale velocity data, all without a seasonal bias. The Argo array has been deployed through the collaboration of more than 40 countries plus the European Union.

A guiding principle of Argo is that the program should benefit everyone, thus the data are openly and immediately available to anyone wishing to use them. Argo data coupled with global-scale satellite measurements from radar altimeters has made possible huge advances in the representation of the oceans in coupled ocean-atmosphere models used for climate forecasts and the routine analysis and forecasting of the state of the subsurface ocean. Argo data are being used in an ever-widening range of research applications that have led to new insights into how the ocean and atmosphere interact in extreme as well as normal conditions. Two examples are the processes in polar winters when the deep waters that fill most of the ocean basins are formed, and the transfer of heat and water to the atmosphere beneath tropical cyclones. Both conditions are crucial to global weather and climate and could not be observed by ships.

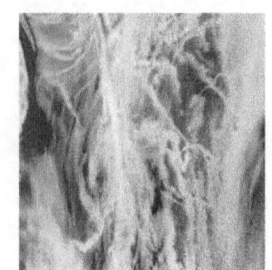

Global Network of Automated Surface-Based Aerosol Measurements.[1] The Aerosol Robotic Network (AERONET) continued expansion of monitoring coverage of the optical properties of atmospheric aerosols (pollution, smoke, desert dust) to a global network of approximately 230 automated sites. Key additional sites were added on the Tibet plateau and in the Ganges floodplain of India to monitor the potential impacts of atmospheric particulates on the region, including possible impacts on atmospheric circulation. New data analysis techniques have been developed that provide a more accurate measure of particle size. In addition, the determination of aerosol light absorption has been improved, which is critical for reducing the current large uncertainties in aerosol radiative forcing of climate. Analysis of data collected in 2004 in the United Arab Emirates with these newly enhanced techniques demonstrated this improved ability to measure aerosol absorption, which leads to a better understanding of the dynamics of desert dust and pollution aerosols over a variety of environments, including Arabian Peninsula desert and over the Persian Gulf. AERONET has also recently included measurements made on ships of opportunity with hand-operated instruments, to better characterize marine environments where no islands exist for automated monitoring. From June through August 2007, AERONET organized numerous ground-based measurements under the CALIPSO satellite flight track in order to validate the satellite products and also to make use of the combined information from both upward- and downward-viewing remote-sensing measurements.

Surface-Based Micro Pulse Lidar Network.[2,3,4] The Micro Pulse Lidar Network (MPLNET) is a federated network of MPL systems designed to measure aerosol and cloud vertical structure continuously, day and night, over long time periods required to contribute to climate change studies and provide validation for models and satellite sensors in NASA's Earth Observing System (see <mplnet.gsfc.nasa.gov>). At present, 13 permanent sites are operational worldwide, with five more to be completed soon. Numerous temporary sites have been established in support of various field campaigns. Most MPLNET sites are co-located with AERONET sun photometer sites to provide both column and vertically resolved aerosol and cloud data, such as optical depth, absorption, size distribution, aerosol and cloud heights, and planetary boundary layer structure and evolution. Recent MPLNET accomplishments include contributions to the development of a novel approach to retrieve the height and optical depth of low, thick cloud layers (such as stratus). Such clouds can contain vast amounts of water and reflect significant amounts of sunlight. However, these clouds are extremely difficult to analyze from space due to their low altitude and high drop concentration (opacity). In another recent study, MPLNET contributed to the most comprehensive assessment of aerosol

profiling capability to date. The study concluded that measured aerosol extinction profile uncertainty is approximately 20% on average. The profile of aerosol extinction is used to determine aerosol radiative effects. The accuracy with which researchers can estimate aerosol extinction directly affects ability to quantify aerosol impacts on climate.

Multi-Platform Field Experiment to Study Tropical Clouds and Climate. **CCSP scientists** completed the Tropical Composition, Cloud, and Climate Coupling (TC4) field experiment in Costa Rica (July to August 2007), which focused on identifying and quantifying chemical and dynamical processes occurring in the tropical tropopause layer. This region of the Earth's atmosphere plays a key role in both climate change science and atmospheric ozone depletion. One of the specific goals of TC4 was to study the composition, formation, and radiative properties of clouds (cirrus and sub-visible cirrus) in this region, thereby assessing the contributions of such clouds, aerosols, and water vapor to climate forcing. Other aspects of the campaign focused on understanding the convective processes that control the transport of air from the lower atmosphere into the tropical tropopause layer (the coldest layer of the atmosphere, at 14 to 18 km altitude) and thence into the stratosphere where they can influence stratospheric ozone. This campaign combined the unique observations from the A-Train satellites and three ground-based and balloon sonde stations in the inter-tropical convergence zone together with three instrumented aircraft flying in a stacked formation (NASA's DC-8, WB-57, and ER-2). Through such coordinated measurements, the TC4 campaign not only sought to address processes controlling the composition of the upper troposphere but also to validate and enhance satellite data analysis. The ER-2 contained downward-looking remote-sensing instruments and flew at an altitude of 20 km, similar to some satellite instruments. The WB-57 contained *in situ* instruments and flew within the upper-troposphere cloud layers to characterize the composition simultaneously observed remotely from above. The DC-8 contained both *in situ* and remote-sensing instruments to quantify the composition of gases, aerosols, and clouds below and into the upper-atmosphere cloud layers transported upward by convection. Numerous research efforts using TC4 data will improve understanding of this important atmospheric region.

CloudSat Measurements. **Data from the CloudSat radar have provided scientists with new** insights into clouds and their structure, and have also provided entirely new insights into Earth's most vital source of freshwater and revealed fascinating views of the massive weather systems that form and die as they circle Earth. Some of the new discoveries offered over the first 12 months of CloudSat operations include:

- The first real information on the fraction of clouds that produce precipitation. These observations indicate that almost 15% of clouds over oceans produce precipitation that reaches the surface. Thus, CloudSat has shown precipitation to be much more common than previously thought, because precipitation over oceans is extremely

hard to measure and light rain that often falls has been completely missed by satellite observations until now. Weather and climate models currently fail to predict this precipitation and it is expected that CloudSat observations will lead to direct improvements in these predictions.

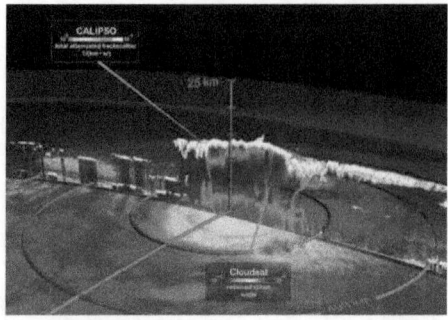

- For the first time, the ability to peer inside major damaging storms, revealing cloud structures and rainfall of hurricanes for the very first time. These observations have provided scientists with new satellite-based estimates of hurricane intensity. These observations are important as they uniquely test theories that shape understanding of how storm intensity might change due to climate change.
- Weather and climate-prediction models indicate that the majority of rainfall comes from deep thunderstorms. However, CloudSat observations now show that a large proportion of rain falls from much shallower clouds.
- New insights on the greenhouse effects of clouds, identifying where and when clouds trap heat in the atmosphere and increase heat lost from the atmosphere to space. This dynamic tradeoff between heating and cooling is one of the basic controls on global climate and this new knowledge gives scientists better tools with which to predict future climate.
- Observations of clouds over polar regions during winter. These clouds have been largely invisible to earlier satellites because of the lack of sunlight and the difficulty of sensing a difference between cold clouds and cold ice-covered surfaces. The polar regions are extremely sensitive to climate warming and the complex interplay between the polar surface and polar clouds can now be studied for the first time.

Cloud-Aerosol Lidar and Infrared Pathfinder Satellite Observation (CALIPSO). **Like CloudSat, the** CALIPSO mission celebrated its first year of operation in June 2007, and is providing new observations of the global distribution and vertical structure of aerosols and thin clouds with unprecedented detail. CALIPSO's innovative measurement capabilities, together with those from the A-Train satellite constellation (see Figure 20), are helping to better understand how aerosols modify Earth's climate by cooling the surface or warming the atmosphere; how they affect cloud lifetimes and precipitation; and how they are lofted into the free troposphere, transported long distances, and affect air quality. CALIPSO's measurements of thin tropical clouds are also providing new insight into processes that maintain the humidity distribution in the upper troposphere. In addition,

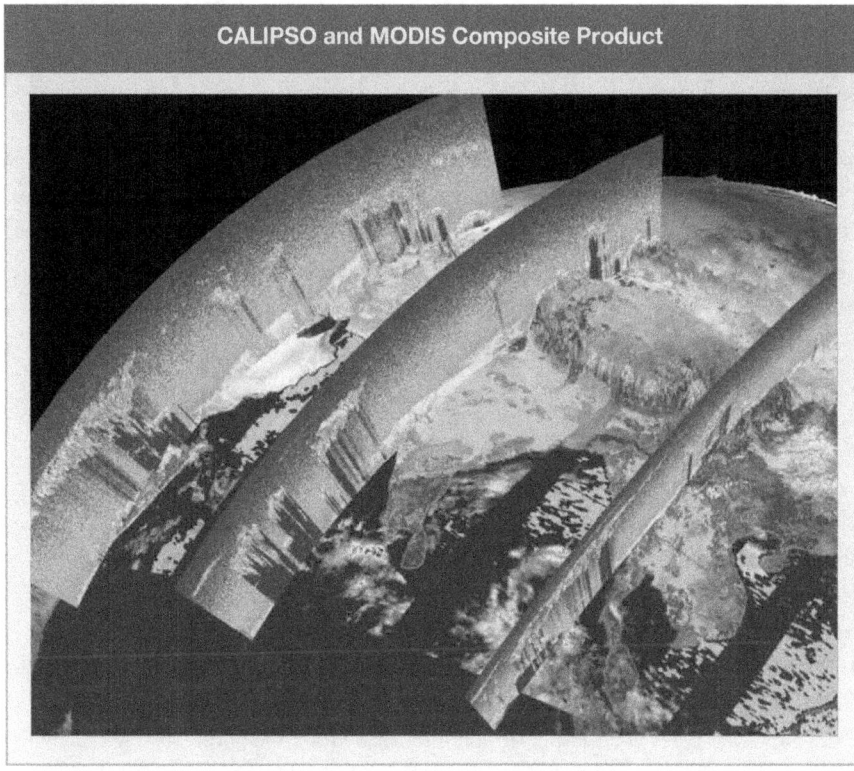

CALIPSO and MODIS Composite Product

Figure 20: CALIPSO and MODIS Composite Product. A composite image showing CALIPSO observations of lidar backscatter superimposed with aerosol optical measurements from Aqua MODIS over south central Asia on 25 October 2006. CALIPSO measurements show aerosols pooling next to the Himalayan mountains (seen by bright yellow and red colors). These data aid studies of aerosols over India and the aerosol-induced warming of the atmosphere, which contribute to retreat of Himalayan glaciers. *Credit: C.R. Trepte, NASA / Langley Research Center.*

these data are being combined with observations from CloudSat to produce a composite survey of the vertical distribution of clouds of varying thickness and layering, especially in the difficult-to-measure polar night. In concert with observations from the Aura satellite, CALIPSO is further aiding in understanding of changes in atmospheric composition by providing new perspectives on the formation and evolution of polar stratospheric clouds that play a key role in the development of the ozone hole.

Surface-Based Observatories of Clouds and Radiation.[5,6,7,8,9,10,11] The Atmospheric Radiation Measurement (ARM) Climate Research Facility (ACRF) provides the infrastructure needed for studies investigating atmospheric processes and for climate model development and evaluation. The ACRF observation resources consist of several highly instrumented stationary facilities, a mobile facility, and aerial vehicles for studying cloud formation processes and their influence on radiative transfer, and for measuring other parameters that determine the radiative properties of the atmosphere. The stationary sites provide scientific test beds in three climatically significant regions (mid-latitude, polar, and tropical), and the mobile facility provides a capability to address high-priority scientific questions in regions not covered by the stationary sites. The aerial vehicles provide a capability to obtain *in situ* cloud and radiation measurements that complement the ground measurements. The ACRF data archive is available to the atmospheric community for climate research in near-real-time. In 2007, the mobile facility was deployed in the Black Forest region of Germany, where scientists studied rainfall resulting from atmospheric uplift (convection) in mountainous terrain (orographic precipitation). Coordinated observations using combinations of mobile and aerial facilities at the ARM fixed sites in the tropics and the Arctic have provided a rich source of information on processes in these regions. Research results have recently been published using these Tropical Warm Pool-International Cloud Experiment and Mixed-Phase Arctic Cloud Experiment coordinated observations. In 2008, the mobile facility will be deployed to China to examine aerosol indirect effects.

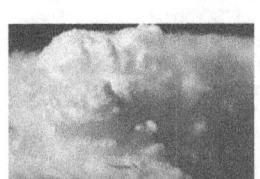

Testing Cloud Models by Cloud Type and Atmospheric Conditions.[12,13] Accurate modeling of clouds in climate prediction models remains the largest uncertainty in climate sensitivity over the next century. Typical climate data sets use monthly mean observations at a 100-km spatial scale to reach sufficient sampling for climate accuracy and then test the ability of climate models to reproduce the monthly gridded observations over the globe. A major limitation in this approach is the inability to relate cause and effect in fast climate processes like clouds. During a month of weather in any particular 100-km grid box on the Earth, many different types of clouds, surface, and atmospheric conditions will have occurred, confounding the ability to decide which clouds need fixing, for what types of atmospheric conditions, and for which processes. Meanwhile, typical field experiments can obtain only a few carefully chosen case studies that have insufficient sampling to test models at climate accuracy. New approaches have been developed to obtain the specificity of field experiments by cloud type and/or atmospheric conditions using global satellite observations such as Clouds and the Earth's Radiant Energy System (CERES), Moderate Resolution Imaging Spectrometer (MODIS), and International Satellite Cloud Climatology Project (ISCCP) combined with global weather data. Early results from these new studies used more than 10,000 cloud systems to study what happens when clouds change on climate time scales such as an El Niño event. The results showed that the physical properties of each cloud type (stratus, cumulus, cumulonimbus) remained remarkably stable, but that the frequency of occurrence of each cloud type changed.

Observing Mass Distribution Changes from Space. The Gravity Recovery and Climate Experiment (GRACE) is a two-spacecraft tandem mission, developed under a partnership between NASA and the Deutsches Zentrum für Luft- und Raumfahrt (DLR) of Germany. After 5 successful years of mission operation, many significant multidisciplinary results using GRACE observations have been reported. The unprecedented accuracy of the measurements provides the opportunity to observe time variability in the Earth's gravity field due to changes in mass distribution. The

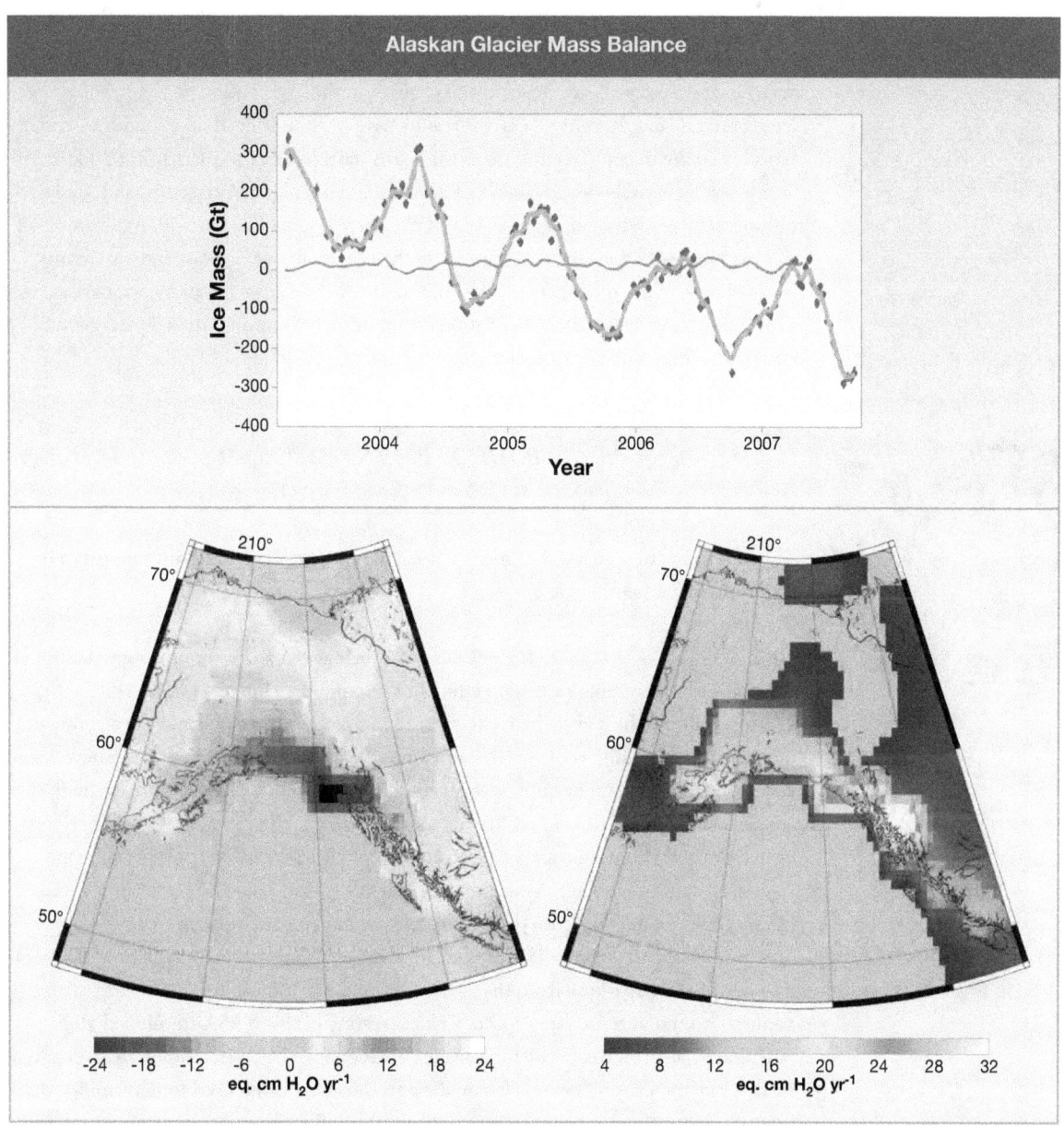

Figure 21: Alaskan Glacier Mass Balance. Alaskan glacier mass balance is provided by analyses of measurements from the GRACE mission's inter-satellite range-rate data (April 2003 through September 2007). The top panel shows the annual variation and overall trend in ice mass for the blue-colored areas in the lower left image (ice mass units are Gt, i.e., 10^{15} g). The lower left image presents the spatial distribution of the surface ice mass trends for the period April 2003 to March 2007. The lower right image shows the spatial distribution of surface annual ice mass change. The greatest negative net balance rates are found in the Yakutat and Glacier Bay regions, while the largest annual amplitudes are found in the southeast glacier regions. Note that surface mass variation contributions from the atmosphere, oceans, tides, terrestrial water storage, and glacial isostatic adjustment have been removed. *Credit: S.B. Luthcke, NASA / Goddard Space Flight Center.*

month-to-month gravity variations obtained from GRACE provide information about changes in the distribution of mass within the Earth and at its surface. The largest time variable gravity signals are the result of changes in the distribution of water, snow, and ice stored on land. Recently GRACE results have been produced on a month-to-month basis for all the major glacier areas of the Earth with areas as small as <50,000 km^2. These results have shown the losses of ice mass in Greenland, Antarctica, and Alaska to be consistent with the observed sea-level rise for the same time period. An example of this exciting work appears as Figure 21 on the previous page. Precise measurements made from satellite orbit may be used to monitor large ice sheets and glacier areas, providing glacier mass balance variations at monthly resolution. GRACE data have emerged as one of the critical climate observations provided by CCSP.

HIGHLIGHTS OF RECENT RESEARCH— DATA MANAGEMENT AND INFORMATION

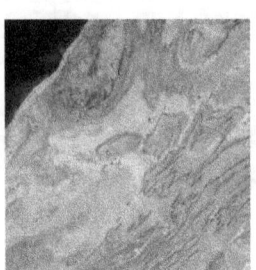

The following are selected data management and information activities supported by CCSP-participating agencies.

Data Fusion for Climate Sensitivity.[14,15] One of the prime challenges in understanding climate is to unscramble cause and effect in this complex climate system. The Earth's energy balance of heat absorbed and emitted can be changed by changes in any one or more of a range of different surface and atmospheric components including snow, ice, vegetation, cloud, temperature, water vapor, carbon dioxide (CO_2), methane, and aerosols. Recent advances in CCSP global satellite instruments have allowed merging of these many different properties into consistent climate-quality data sets allowing analysis of cause and effect in changing the Earth's energy balance. For example, the NASA CERES energy balance data products now merge data from up to 11 instruments on seven spacecraft from NASA, NOAA, and international partners. The fused CERES data describe energy flow from the surface through the atmosphere to the top of atmosphere and out into deep space. Critical testing of this NASA-developed global radiation balance data is performed against a wide range of DOE- and NOAA-operated surface radiation sites. The new fusion data products are being used to determine what part of climate changes over the course of Earth's reflectance of solar energy back to space are caused by clouds or by snow and ice—both key feedbacks that determine climate sensitivity.

Carbon Dioxide Information Analysis Center (CDIAC). DOE's CDIAC provides comprehensive, long-term data management support, analysis, and information services to DOE's climate change research programs, the global climate research community, and the

general public. The CDIAC data collection is designed to answer questions pertinent to both the present-day carbon budget and temporal changes in carbon sources and sinks. The data sets provide quantitative estimates of anthropogenic CO_2 emission rates, atmospheric concentration levels, land-atmosphere fluxes, ocean-atmosphere fluxes, and oceanic concentrations and inventories. In 2008, CDIAC will augment its ocean holdings by offering CO_2 measurements from buoys, research cruises, and volunteer observing ship lines along U.S. coastlines to support the North American Carbon Program (NACP). In 2008, CDIAC will also release the final Carbon Dioxide in the Atlantic Ocean (CARINA) synthesis database including both discrete and underway measurements. CDIAC will release the final North Pacific Marine Science Organization (PICES) synthesis database, which will replace the previous North Pacific discrete measurement component of the Global Ocean Data Analysis Project.

Quality Assurance for the Global Atmosphere Watch (GAW) Precipitation Chemistry Program. Precipitation chemistry remains a major environmental issue due to concerns over eutrophication, ecosystem health, biogeochemical cycling, and global climate change. Although global modeling assessments require data of high and known quality, many of the laboratories supporting the approximately 200-site global network require expert assistance and ongoing oversight. CCSP agency scientists—in close cooperation with the State University of New York at Albany, Environment Canada, and European, East Asian, and other scientists—has addressed these problems through the development and provision of a guidance manual for program participants, and the development of a tool for rapid assessment of laboratory quality by data users. Intercomparisons have been conducted annually since in 1985 and biannually since 2001.[a] In addition to complete quality assurance information, it is the goal of this program to make all GAW precipitation chemistry data freely downloadable from the Internet.

Global Observing System Information Center. The Global Observing System Information Center (GOSIC) began as a developmental activity at the University of Delaware in 1997, and as of January 2007 had been fully converted to an operational global data facility through CCSP agency support on behalf of and with the concurrence of the global observing community. GOSIC provides information, and facilitates easier access to data and information produced by GCOS, the Global Ocean Observing System (GOOS), the Global Terrestrial Observing System (GTOS), and their partner programs. The distributed nature of this vast system of global and regional data and information systems is best served by such a single entry point for users. GOSIC provides explanations of the various global data systems, as well as providing an integrated overview of the

---

[a] Global laboratory intercomparison data are presently posted at <qasac-americas.org> and may be displayed by clicking on the "Data" link and then on "Ring Diagram Assessments."

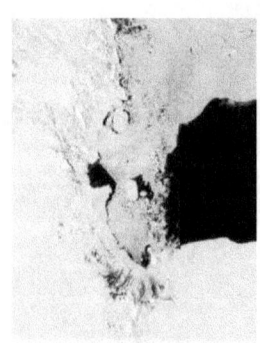

various global observing programs, which includes on-line access to data, information, and services. GOSIC offers a search capability across international data centers in order to better facilitate access to a worldwide set of observations and derived products. See <gosic.org> for more detail.

Annual State of the Climate Report—Using Earth Observations to Monitor the Global Climate. In partnership with WMO, along with numerous national and international partners, a State of the Climate monitoring effort has been established, which consists of operational monitoring, analysis, and reporting on atmosphere, ocean, and land surface conditions from the global to local scale. By combining historical data with current observations, this program places present-day climate in historical context and provides perspectives on the extent to which the climate continues to vary and change as well as the effect that climate is having on societies and the environment. More than 150 scientists from over 30 countries are now part of an annual process of turning raw observations collected from the global array of observing systems into information that enhances the ability of decisionmakers to understand the state of the Earth's climate and its variation and change during the past year, with context provided by decades to centuries of climate information. Many observing and analysis systems are unique to countries or regions of the world, but through this effort, the information from each system is openly shared and has proven essential to moving data into operational use and filling critical gaps in current knowledge about the state of the global climate system (see <ncdc.noaa.gov/oa/climate/research/state-of-climate> for more detail). A State of the Climate report is distributed through publication in the *Bulletin of the American Meteorological Society* each year. Working with WMO, this report is also translated into other languages and distributed to all 187 WMO member nations. The State of the Climate Report seeks to report on as many of the Essential Climate Variables as possible as identified by the GCOS Second Adequacy Report.

Polar Ice Albedo and Cloud Feedback.[15,16,17] The International Polar Year is underway in 2007 and 2008. One of the key elements to understand is the role of snow and ice albedo feedback in amplifying the sensitivity of climate. As snow and ice retreat in a warming climate, they expose darker, less reflective surfaces that can allow additional absorption of solar radiation, and therefore further amplify polar and global warming. New global satellite observations since 2000 allow the data fusion of climate accuracy snow data (MODIS), sea-ice cover [Advanced Microwave Scanning Radiometer (AMSR)], cloud properties (MODIS), and global albedo/reflectance (CERES) to study this key feedback with an accuracy never before available. A key advance has been the ability of MODIS to derive more accurate satellite-measured polar snow cover and clouds verified against the DOE Barrow Alaska surface site, and the ability of CERES to derive more accurate polar reflectance to space as a function of snow and cloud changes as well as

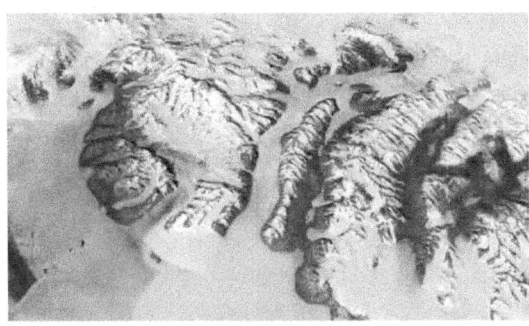

new estimates of surface
radiation verified against the
U.S. Baseline Surface Radiation
Network (BSRN) and DOE
ARM surface sites. The new
data confirm that snow and ice
retreat are significantly
increasing Earth's absorption of
solar energy by reducing its
reflectance to space. These new data also include the darkening effects of increasing
vegetation in polar regions. At the same time that clear-sky conditions show lower
Earth reflectance (warming effect), the data show that much of the drop in reflectance
of the Earth due to snow and ice retreat has been offset by an increased reflectance
from increasing cloud cover during polar summer (cooling effect). But these same
clouds can act to reduce the polar surface emission of thermal radiation to space
(especially in polar winter), and further research will look at the total effect of solar
and thermal infrared energy in summer and winter seasons.

Highlights of Plans for FY 2009

CCSP will continue to develop and implement integrated systems for observing and
monitoring global change, and the associated data management and information
systems. Selected key planned activities for FY 2009 and beyond follow.

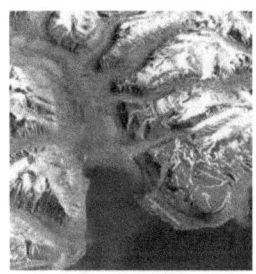

Global Climate and Ocean Observing Systems. FY 2009 priorities for advancing the
atmospheric and ocean observing components of GCOS include: (1) reducing the
uncertainty in the carbon inventory of the global ocean, sea-level change, and sea
surface temperature; (2) continuing support for existing *in situ* atmospheric networks
in developing nations; and (3) planning for surface and upper air GCOS reference
observations consistent with CCSP Synthesis and Assessment Product 1.1. As such, the
global ocean observing system will make incremental advances, building up to 62%
completion: 50 surface drifters will be equipped with salinity sensors for satellite
validation and salinity budget calculations, particularly in the polar regions; a new
reference array will be added across the Atlantic basin to measure changes in the
ocean's overturning circulation, an indicator of possible abrupt climate change; a pilot
U.S. coastal carbon observing network will enter sustained service to help quantify
North American carbon sources and sinks and to measure ocean acidification caused by
CO_2 sequestration in the ocean; and dedicated ships will be deployed to target
deployments of Argo and surface drifters in undersampled regions of the world oceans.

Finally, planning activities will continue on developing a GCOS Reference Upper Air Network (GRUAN) to aid in enhancing the quality of upper tropospheric and lower stratospheric water vapor measurements at a subset of present GCOS Upper Air Network stations.

These activities will address Goals 12.3 and 12.5 of the CCSP Strategic Plan.

Extended Examination and Intercomparison of Water Vapor Measurements from Aircraft, Balloons, and Satellites. Water vapor is the most important greenhouse gas in the atmosphere, exhibiting large gradients in concentration and mixing ratio between the Earth's surface and the upper troposphere/lower stratosphere (UT/LS). Fitting in with the GRUAN planning work, understanding changes in the distribution of water vapor, whether due to natural or anthropogenic causes, is essential to understanding the potential for climate change. Even small increases in stratospheric water vapor (1% per year) could cause significant surface radiative forcing and stratospheric cooling. Stratospheric water vapor amounts are controlled by dehydration processes driven by low temperatures in the tropopause region of the tropics. Understanding of the dehydration process and its variability is incomplete. Of particular importance is the extent and frequency of ice-supersaturated conditions in the UT/LS. These shortfalls in knowledge have made accurate and precise water vapor measurements in the tropopause region a required component of future climate research, particularly at the low water vapor mixing ratios in the UT/LS where measurement discrepancies currently exist. A number of research efforts will be continued or initiated to help resolve the observed discrepancies in *in situ* water vapor observations. CCSP agencies are jointly conducting these activities with the involvement of U.S. and international investigators from a wide range of government and academic institutions. The planned efforts include: (1) single instrument laboratory studies designed to better characterize and understand instrument performance and calibration under a variety of atmospheric conditions; (2) the possible selection and use of a water vapor calibration standard to establish and/or confirm measurement accuracy and precision; and (3) multiple-instrument intercomparisons in the laboratory and field involving an independent referee to coordinate and present the results of each formal laboratory and flight intercomparison that includes instruments from different research groups. Field intercomparisons will include aircraft-, balloon-, and satellite-borne instruments.

These activities will address Goals 12.3 and 12.5 of the CCSP Strategic Plan.

International Polar Year Observations. The United States will conduct aircraft flights over the North Slope of Alaska to measure temperature, humidity, total particle number, aerosol size distribution, cloud condensation nuclei concentration, ice nuclei concentration, optical scattering and absorption, vertical velocity, cloud liquid water and ice contents, cloud droplet and crystal size distributions, cloud particle shape, and

cloud extinction. These data, coupled with ground-based measurements, will be used to evaluate model simulations of Arctic climate. The CALIPSO lidar and CloudSat radar are providing satellite measurements of the difficult-to-observe polar clouds. The last of these capabilities will also directly support IPY activities.

CCSP researchers will begin analysis of data from a series of FY 2008 airborne field campaigns addressing Arctic climate. These analyses of data from aircraft flights, ground measurements, and satellites will contribute to a larger international effort called POLARCAT (Polar Study using Aircraft, Remote Sensing, Surface Measurements, and Models of Climate, Chemical Aerosols, and Transport). Spring observations will be analyzed to assess the long-range transport of anthropogenic pollution to the Arctic and its contribution to Arctic haze and tropospheric ozone chemistry. Summer observations will be analyzed to assess boreal fire emissions. These analyses will ultimately improve the ability of current models to simulate the influence of anthropogenic pollution and boreal fires on the Arctic atmosphere and climate as it relates to changing atmospheric composition, radiative forcing of trace gases and aerosols, and aerosol-cloud interactions.

Finally, two U.S. Climate Reference Network systems will be deployed at the Russian arctic sites of Tiksi (72.5°N) and Yakutsk (63.0°N) in order to provide long-term reference measurements of temperature, precipitation, wind, pressure, and surface radiation in support of IPY and beyond.

These activities will address Goals 12.3 and 12.5 of the CCSP Strategic Plan.

Surface-Based Measurements of Aerosols and Clouds. AERONET retrievals of atmospheric particulate absorption will continue to be utilized in climate forcing studies and in the validation of current and future satellite missions, such as the Glory satellite (early 2009 launch), which will measure aerosol light absorption from space. Network expansion will continue, with a focus on inadequately sampled regions that are important for understanding global climate change, such as China (both the polluted eastern regions and the western deserts that are a source of dust storms). An experimental effort is underway to investigate the possibility of measuring sunlight reflected off the moon to make aerosol measurements at night. In addition, an experimental algorithm is under development to make measurements of atmospheric CO_2.

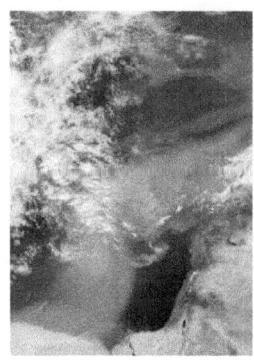

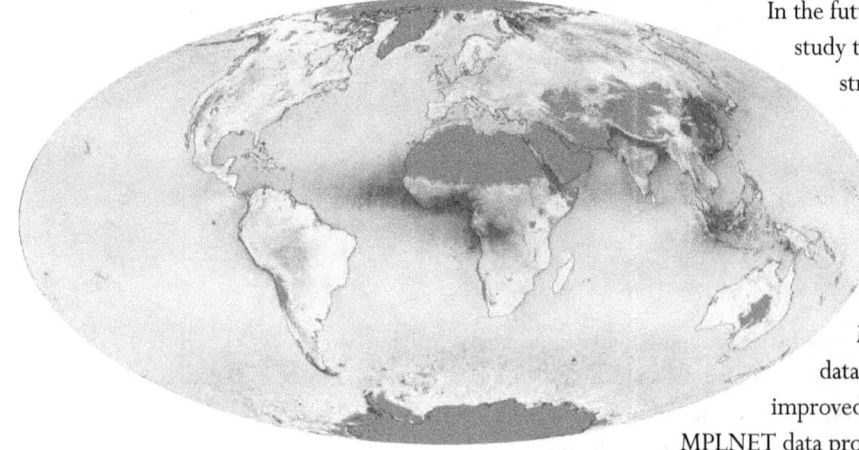

In the future, lidar data will be used to study the influence of polar stratospheric clouds on ozone formation over the South Pole, to study Arctic haze impacts on polar climate, and to generate climatological aerosol and cloud properties at several MPLNET sites. To enhance data value, MPL instrument improved. In addition, several new MPLNET data products will be made available to the research community.

These activities will address Goals 12.1 and 12.5 of the CCSP Strategic Plan.

Solar Variability: Glory. The Glory mission is planned to launch in 2009. It will carry a Total Irradiance Monitor (TIM) based on the Solar Radiation and Climate Experiment (SORCE) TIM design, with the same high-precision phase-sensitive detection capability. Glory will also carry an Aerosol Polarimeter Sensor (APS), which will improve the ability to distinguish among aerosol types by measuring the polarization state of reflected sunlight. Both TIM and APS will provide key measurements beginning in 2009 during the minimum of solar cycle 24. This less-active portion of the 11-year solar cycle is especially crucial in estimating any long-term trends in solar output—a key to understanding the 20th-century context of global change, as the Sun is the single entirely "external" forcing of the climate system that is unaffected by climate change itself.

These activities will address Goals 12.1 and 12.5 of the CCSP Strategic Plan.

Global Precipitation Measurement Mission. Motivated by the successes of the Tropical Rainfall Measuring Mission (TRMM) satellite and recognizing the need for a more comprehensive global precipitation measuring program, NASA and the Japan Aerospace Exploration Agency conceived a new Global Precipitation Measurement (GPM) mission

that is still in the formulation phase. A fundamental scientific goal of GPM is to make substantial improvements in global precipitation observations, especially in terms of measurement accuracy, sampling frequency, spatial resolution, and coverage, thus extending TRMM's rainfall time series. To achieve this goal, the mission will consist of a constellation of low-Earth-orbiting satellites carrying various passive and active microwave measuring instruments. The GPM mission will be used to address important issues central to improving the predictions of climate, weather, and hydrometeorological processes; to stimulate operational forecasting; and to underwrite an effective public outreach and education program, including near-real-time dissemination of televised regional and global rainfall maps. Assessment of how natural and anthropogenic aerosols affect precipitation variability (and therefore the water cycle) is a complex and important problem. The capability to monitor the diurnal cycle of rainfall globally with GPM is expected to enable significantly improved understanding of the links between aerosols, climate variability, weather changes, hydrometeorological anomalies, and small-scale cloud macrophysics and microphysics.

These activities will address Goals 12.1 and 12.5 of the CCSP Strategic Plan.

Aquarius. Aquarius is a satellite mission to measure global sea surface salinity. Its instruments will measure changes in sea surface salinity over the global oceans to a precision of 2 parts in 10,000 (equivalent to about 1/6 of a teaspoon of salt in 1 gallon of water). By measuring global sea surface salinity with good spatial and temporal resolution, Aquarius will answer long-standing questions about how the oceans respond to climate change and the water cycle, including changes in freshwater input and output to the ocean associated with precipitation, evaporation, ice melting, and river runoff. Aquarius is a collaboration between NASA and the Comison Nacional de Actividades Espaciales (CONAE), the Argentine space agency, with an expected launch date in 2009.

These activities will address Goals 12.1 and 12.5 of the CCSP Strategic Plan.

Ocean Surface Topography Mission. The accurate, climate-quality record of sea surface topography measurements, started in 1992 with TOPography EXperiment (TOPEX)/Poseidon and continued in 2001 by the Jason satellite mission, will be extended with the Ocean Surface Topography Mission (OSTM). These missions have provided accurate estimates of regional sea-level change and global sea-level rise unbiased by the uneven distribution of tide gauges. Ocean topography measurements from these missions have elucidated the role of tides in ocean mixing and maintaining deep ocean circulation. Further, quantitative determination of ocean heat storage from satellite measurements together with measurements from the Argo global array of temperature/salinity profiling floats have confirmed climate model predictions of the Earth's energy imbalance that is primarily due to greenhouse gas forcing. The high levels of absolute accuracy and cross calibration make these missions uniquely suited for climate

research. OSTM is a collaboration among NASA, NOAA, the French space agency Centre National d'Etudes Spatiales (CNES), and the European Organisation for the Exploitation of Meteorological Satellites (EUMETSAT).

These activities will address Goals 12.1 and 12.5 of the CCSP Strategic Plan.

Orbiting Carbon Observatory. The Orbiting Carbon Observatory (OCO) is a new mission, expected to launch in 2008, that will provide the first dedicated, space-based measurements of atmospheric CO_2 (total column) with the precision, resolution, and coverage needed to characterize carbon sources and sinks on regional scales and to quantify their variability. Analyses of OCO data will regularly produce precise global maps of CO_2 in the Earth's atmosphere that will enable more reliable projections of future changes in the abundance and distribution of atmospheric CO_2 and studies of the effect that these changes may have on Earth's climate.

These activities will address Goals 12.2 and 12.5 of the CCSP Strategic Plan.

Integrated Ocean Observing System (IOOS). IOOS is the U.S. coastal observing component of GOOS and is envisioned as a coordinated national and international network of observations, data management, and analyses that systematically acquires and disseminates data and information on past, present, and future states of the oceans. A coordinated IOOS effort has been established by CCSP via a national IOOS Program Office co-located with the <Ocean.US> consortium of offices consisting of NASA, NSF, NOAA, and the Navy (see <ocean.us>). The IOOS observing subsystem employs both remote and *in situ* sensing. Remote sensing includes satellite-, aircraft- and land-based sensors, power sources, and transmitters. *In situ* sensing includes platforms (ships, buoys, gliders, etc.), *in situ* sensors, power sources, sampling devices, laboratory-based measurements, and transmitters.

These activities will address Goals 12.3 and 12.6 of the CCSP Strategic Plan.

OBSERVING AND MONITORING THE CLIMATE SYSTEM
CHAPTER REFERENCES

1) **Eck**, T.F., B.N. Holben, J.S. Reid, A. Sinyuk, O. Dubovik, A. Smirnov, D. Giles, N.T. O'Neill, S.-C. Tsay, Q. Ji, A. Al Mandoos, M. Ramzan Khan, E.A. Reid, J.S. Schafer, M. Sorokine, W. Newcomb, and I. Slutsker, 2008: Spatial and temporal variability of column-integrated aerosol optical properties in the southern Arabian Gulf and United Arab Emirates in summer. *Journal of Geophysical Research*, 113, D01204, doi:10.1029/2007JD008944.

2) **Welton**, E.J., J.R. Campbell, J.D. Spinhirne, and V.S. Scott, 2001: Global monitoring of clouds and aerosols using a network of micro-pulse lidar systems. In: *Lidar Remote Sensing for Industry and Environmental Monitoring* [Singh, U.N., T. Itabe, and N. Sugimoto (eds.)]. *Proceedings of SPIE*, **4153**, 151-158.

OBSERVING AND MONITORING THE CLIMATE SYSTEM
CHAPTER REFERENCES (CONTINUED)

3) **Chiu**, J.C., A. Marshak, W.J. Wiscombe, S.C. Valencia, and E.J. Welton, 2007: Cloud optical depth retrievals from solar background signals of micropulse lidars. *IEEE Geosciences and Remote Sensing Letters*, **4**, 456-460.

4) **Schmid**, B., R. Ferrare, C. Flynn, R. Elleman, D. Covert, A. Strawa, E. Welton, D. Turner, H. Jonsson, J. Redemann, J. Eilers, K. Ricci, A.G. Hallar, M. Clayton, J. Michalsky, A. Smirnov, B. Holben, and J. Barnard, 2006: How well do state-of-the-art techniques measuring the vertical profile of tropospheric aerosol extinction compare? *Journal of Geophysical Research*, **111**, D05S07, doi:10.1029/2005JD005837.

5) **McFarlane**, S.A. and W.W. Grabowski, 2007: Optical properties of shallow tropical cumuli derived from ARM ground-based remote sensing. *Geophysical Research Letters*, **34**, L06808, doi:10.1029/2006GL028767.

6) **McFarlane**, S.A, J.H. Mather, and T.P. Ackerman, 2007: Analysis of tropical radiative heating profiles: A comparison of models and observations. *Journal of Geophysical Research*, **112**, D14218, doi:10.1029/2006JD008290.

7) **Prenni**, A.J., J.Y. Harrington, M. Tjernstrom, P.J. Demott, A. Avramov, C.N. Long, S.M. Kreidenweis, P.Q. Olsson, and J. Verlinde, 2007: Can ice-nucleating aerosols affect arctic seasonal climate? *Bulletin of the American Meteorological Society*, **88(4)**, 541-550.

8) **Lubin**, D. and A.M. Vogelmann, 2007: Expected magnitude of the aerosol shortwave indirect effect in springtime Arctic liquid water clouds. *Geophysical Research Letters*, **34(11)**, L11801, doi:10.1029/2006GL028750.

9) **Hume**, T. and C. Jakob, 2007: Ensemble single column model validation in the tropical western Pacific. *Journal of Geophysical Research*, **112(D10)**, D10206, doi:10.1029/2006JD008018.

10) **McFarquhar**, G.M., J. Um, M. Freer, D. Baumgardner, G.L. Kok, and G. Mace, 2007: Importance of small crystals to cirrus properties: Observations from the Tropical Warm Pool International Cloud Experiment (TWP-ICE). *Geophysical Research Letters*, **34**, L13803, doi:10.1029/2007GL029865.

11) **McFarquhar**, G.M., G. Zhang, M.R. Poellot, G.L. Kok, R. McCoy, T. Tooman, A. Fridlind, and A.J. Heymsfield, 2007: Ice properties of single layer stratocumulus during the Mixed-Phase Arctic Cloud Experiment (M-PACE): Part I. Observations. *Journal of Geophysical Research*, **112**, D24201, doi:10.1029/2007JD008633.

12) **IPCC**, 2007: Summary for Policymakers. In: *Climate Change 2007: The Physical Science Basis. Contribution of Working Group I to the Fourth Assessment Report of the Intergovernmental Panel on Climate Change* [Solomon, S., D. Qin, M. Manning, Z. Chen, M. Marquis, K. B. Averyt, M. Tignor and H. L. Miller (eds.)]. Cambridge University Press, Cambridge, United Kingdom and New York, NY, USA, pp. 1-18.

13) **Xu**, K-M., T. Wong, B.A. Wielicki, L. Parker, B. Lin, Z.A. Eltzen, and M. Branson, 2007: Statistical analyses of satellite cloud object data from CERES. *Journal of Climate*, **20**, 819-842.

14) **Loeb**, N.G., B.A. Wielicki, W. Su, K. Loukachine, W. Sun, T. Wong, K.J. Priestley, G. Matthews, W.F. Miller, and R. Davies, 2007: Multi-instrument comparison of top-of-atmosphere reflected solar radiation. *Journal of Climate*, **20(3)**, 575-591.

15) **Kato**, S., N.G. Loeb, P. Minnis, J.A. Francis, T.P. Charlock, D.A. Rutan, E.E. Clothiaux, and S. Sun-Mack, 2006: Seasonal and interannual variations of top-of-atmosphere irradiance and cloud cover over polar regions derived from the CERES data set. *Geophysical Research Letters*, **33**, L19804, doi:10.1029/2006GL026685.

16) **Wang**, X. and J. Key, 2005: Arctic surface, cloud, and radiation properties based on the AVHRR Polar Pathfinder data set. *Journal of Climate*, **18(14)**, 2575-2593.

17) **Francis**, J.A., E. Hunter, J. Key, and X. Wang, 2005: Clues to variability in Arctic minimum sea ice extent. *Geophysical Research Letters*, **32**, L21501, doi:10.1029/2005GL024376.

9 | Communications

In its Strategic Plan, CCSP identified communications as one of four core approaches for achieving its five overarching scientific goals. CCSP is committed to communicating with interested partners in the United States and throughout the world, and to learning from these partners on a continuing basis. As an essential part of its mission, CCSP stresses openness and transparency in its findings and reports.

The Communications Interagency Working Group (CIWG), established during FY 2004, develops and executes an implementation plan each year that focuses on disseminating the results of CCSP activities credibly and effectively and making CCSP science findings and products easily available to a diverse set of audiences. Elements of the implementation plan for calendar year 2008 include:

- *Media Relations* — When requested by the CCSP Director, assist in communicating on matters relating to climate science.
- *Public Outreach* — Develop materials and methods for public outreach on issues related to climate science and the activities and products of CCSP.
- *Web Sites* — Develop and advance a strategy for improving, integrating, and promoting the content of web sites operated or supported by CCSP and its participating agencies, recognizing that the sites are essential communication and outreach tools.

The elements and strategies described above are exercised in the launch of the CCSP Synthesis and Assessment Products (SAPs) listed on page 29 of this report. When the final products are released, the teams associated with each SAP may request the CIWG's assistance in generating outreach products or activities via the lead agency and the lead agency's representative to the CIWG. These products and activities may include:

- Constituent briefings
- Web links posted on participating agency and CCSP home pages
- Press releases
- Public comments package posted on web sites
- Fact sheets
- Scientific posters.

Agencies represented on the CIWG incorporate the results of SAPs into their outreach and education materials. For example, EPA includes key summary information from SAP 1.1 on its climate change web site (see <epa.gov/climatechange/science/recenttc.html#recent>).

The Climate Change Science Program Office (CCSPO), funded and supervised by the agencies and departments participating in CCSP, supports the program's communications goals, along with members of the CIWG. CCSPO assists CIWG, coordinates preparation of the annual *Our Changing Planet* report to Congress as well as other reports, and is responsible for managing the program's interagency web sites.

Highlights of Recent Interagency Communications Activity

Listed below are highlights of recent communication activities coordinated at the interagency level (as of 30 June 2008):

- Published and distributed (in both hardcopy and online) the FY 2008 edition of *Our Changing Planet*, the program's annual report to Congress and the President.
- Produced and distributed an additional six SAP final reports, including a series of briefings and other activities focused on communicating the report findings:
  - SAP 2.2: *North American Carbon Budget and Implications for the Global Carbon Cycle* (November 2007)
  - SAP 3.3: *Weather and Climate Extremes in a Changing Climate: Regions of Focus - North America, Hawaii, Caribbean, and U.S. Pacific Islands* (June 2008)
  - SAP 4.3: *The Effects of Climate Change on Agriculture, Land Resources, Water Resources, and Biodiversity* (May 2008)
  - SAP 4.4: *Preliminary Review of Adaptation Options for Climate-Sensitive Ecosystems and Resources* (June 2008)
  - SAP 4.5: *Effects of Climate Change on Energy Production and Use in the United States* (October 2007)
  - SAP 4.7: *Impacts of Climate Variability and Change on Transportation Systems and Infrastructure: Gulf Coast Study* (March 2008)
- Posted on-line drafts of SAP reports for public comment. All submitted public comments also were posted, along with the authors' responses.
- Posted peer review comments on draft SAP reports, along with the authors' responses.
- Advisory committees for many of the SAPs convened public meetings to discuss report drafts. All meetings were announced in the Federal Register.
- Solicited public comment on the summary of CCSP's *Revised Research Plan*. Launched in May 2008, this plan updates the CCSP Strategic Plan released in 2003.

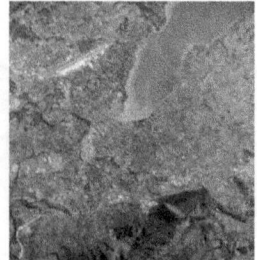

- Facilitated stakeholder participation in the U.S. Government Review of draft documents of the Intergovernmental Panel on Climate Change (IPCC)
- Supported the dissemination of the Fourth Assessment Report of the IPCC, including respective working groups' Summary for Policymakers and the *Synthesis Report*.
- Managed and improved CCSP web sites, including web services to facilitate interagency collaboration.
- Managed the Global Change Research Information Office (GCRIO) as mandated by the Global Change Research Act of 1990, including a catalog for requesting reports.
- Managed the Climate Change Technology Program (CCTP) public web site and provided additional services such as development and management of password-protected web sites and publications support.

Another important set of communication activities by CCSP agencies is a continuing series of workshops and conferences that have been instrumental in conveying recent scientific advances. They have also provided an opportunity for CCSP to better understand stakeholder needs, concerns, and recommendations, and have been used to obtain broad input into CCSP's ongoing strategic planning process, for example:

- *Coping with Climate Change: National Summit, 8-10 May 2007, Ann Arbor, Michigan* — Funded in part by EPA and NSF, the summit focused on four specific sectors that represent illustrative examples of the social, economic, environmental, and natural resource issues that are vulnerable to climate change: public health, the energy industry, water quality, and fisheries. The summit discussed general models for how different kinds of organizations, within these sectors and more generally, can put into place structures or processes that help them to anticipate and adapt to near- and long-term change.
- *Climate Information Users Roundtable, 10 October 2007, Washington, DC* — The purpose of the roundtable was for CCSP to gain insights from a range of stakeholders in order to help inform future program directions. Participants were asked to reflect on their expectations for and ideas about CCSP's role in providing climate information (see <climatescience.gov/Library/docs/ClimateInformationUsersRoundtable-FinalReport.pdf>).
- *Climate Information: Responding to User Needs, 22-23 October 2007, College Park, Maryland* — Funded in part by CCSP agencies, this event fostered dialog between the providers of climate information and its diverse user community to define specific measures needed to enhance data management, modeling, predictions in making climate-related decisions, and the use of climate observations. Providers learned how climate change affects users from different sectors of society, and what specific products they require. Users heard what other organizations are doing to prepare for climate change impacts and what types of information providers can produce now and in the future.

- *NOAA Data and Information for Climate Change: A Conference for Public and Private Sector Users, 5-6 November 2007, Asheville, North Carolina* – This conference focused on identifying recommendations from and requirements of the energy, insurance, and transportation sectors for data and information in a changing climate (see <noaadata.com>). The conference explored the challenges and opportunities that changing climate conditions present for businesses and state and local governments in these sectors; highlighted current scientific understanding of climate change within these sectors; and assessed the energy, insurance, and transportation sectors' emerging data and information needs to better respond to a changing climate.
- *Climate Change: Science and Solutions. National Council for Science and the Environment, 8th National Conference on Science, Policy, and the Environment, 16-18 January 2008, Washington, DC* – Funded in part by seven CCSP agencies, this dialog with leading scientists, policymakers, industry leaders, educators, and other solutions-oriented innovators focused on developing comprehensive strategies for protecting people and the planet from the threat of climate change. A set of recommendations is forthcoming, based upon the discussions in dozens of skill-building workshops, targeted breakout sessions, plenary sessions, and symposia.

HIGHLIGHTS OF PLANS FOR FY 2009

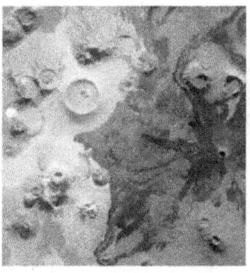

Listed below are some of the communications activities coordinated at the interagency level and planned for late FY 2008 (July to September 2008) and early FY 2009 (October 2008 to May 2009):

- Produce and distribute the balance of 13 SAP final reports (see page 29 for the breadth of topics to be addressed) and a Unified Synthesis Product that will consolidate key SAP findings, recent work of the IPCC, and new peer-reviewed regional climate impact assessment reports. CIWG will advise the lead agencies on communications-related issues including dissemination to appropriate stakeholders, briefings, press releases, and summaries for a range of audiences across different information media.
- Publish and distribute in hardcopy the CCSP *Revised Research Plan*, as specified in Section 104 of the Global Change Research Act of 1990.
- Publish and distribute the *Scientific Assessment of the Effects of Global Change on the United States* (released electronically on 29 May 2008) that addresses topics required in Section 106 of the Global Change Research Act of 1990, including a description of the trends in and projections of changes in the physical climate system, as well as impacts on social and natural adaptation options.
- Prepare and disseminate the FY 2010 edition of *Our Changing Planet*.
- Continue to improve and expand web sites by preparing and posting new content, improving web site usability and accessibility, and enhancing agency integration.

10 | International Research and Cooperation

Global change research, modeling, and observations from institutions based in the United States contribute to and benefit from a number of ongoing international activities. CCSP, the individual agencies that comprise CCSP, its various interagency working groups, and, in particular, the Interagency Working Group on International Research and Cooperation participate in and provide support for a variety of international research activities that collectively cover the broad spectrum of global environmental change research.

Through such active participation and leadership, CCSP and the large community of U.S. scientists supported by or associated with it truly has a global reach. Activities in which the United States is involved include supporting global environmental change research programs including, but not limited to, those that operate under the aegis of the International Council for Science (ICSU); supporting international assessments, particularly the Intergovernmental Panel on Climate Change (IPCC); supporting regional global change research networks; playing an active role in informal international organizations that are involved with the advancement of global environmental change research; and participating in and in many cases leading international efforts to advance coordination and cooperation around observation of the Earth.

Individual CCSP agencies support international activities that are aligned with their goals or missions. In some cases, an agency will be given the lead for a particular effort for the Federal government; this may involve intra- and/or interagency coordination as well as funding, including in-kind support, depending upon the organization. CCSP is also a vehicle for communication and coordination, both within the Federal government and with the broader scientific community, of global change-related information and input to various international organizations. This support includes work with the Department of State at a variety of levels, but particularly with respect to the IPCC and the United Nations Framework Convention on Climate Change (UNFCCC) as well as bilateral arrangements in climate change science and technology.

The United States, through CCSP, also participates actively in informal activities that are dedicated to coordinating and fostering international global environmental change research. One such organization is the International Group of Funding Agencies for Global Change Research (IGFA). IGFA serves as a direct link to the international global change research programs and serves as a way for representatives from CCSP to interact informally with representatives from other countries who have as their responsibility funding of global change research.

CCSP provides the core of the U.S. portion of funding for coordination of international global change research. This includes support for IPCC Working Group I U.S.-hosted Technical Support Unit, which will have completed its work as of 1 September 2008. The Department of State is currently involved in international deliberations regarding selection of host nations for the Working Groups comprising the Fifth Assessment Report. CCSP support is also provided to the partner programs of the Earth System Science Partnership (ESSP) including the SyTem for Analysis, Research, and Training (START). The National Science Foundation on behalf of CCSP manages U.S. support for regional global change research networks, including the Inter-American Institute for Global Change Research (IAI), the Asia-Pacific Network (APN), and the African Network for Earth System Science (AfricanNESS).

The international global change research programs continue to provide sound frameworks for core research projects, capacity building programs, and regional networks. These programs include the World Climate Research Programme (WCRP), the International Geosphere-Biosphere Programme (IGBP), the International Human Dimensions Programme (IHDP), DIVERSITAS (an international biodiversity science program), ESSP, and START. The key regional programs are APN, IAI, AfricanNESS, and several regional programs under the START umbrella (Southeast Asia Regional Centre, Temperate East Asia Regional Committee, etc.). These regional programs, due to their ability to bring together national networks of global change scientists in an international setting, are increasingly being called upon to provide input to international organizations, international assessments, and other activities.

These programs are also highly effective at developing linkages between national networks of scientists, between disciplines, and developing capacity in young scientists and scientists from developing countries. One of the means they utilize is extensive in-person Open Science Conferences (OSCs), congresses, workshops, and other activities.

OSCs bring together the wider science community focused on a specific topic, program, or programs to encourage dialog, connections, communication, and cooperation. They also assist in charting the course for the overarching efforts of the organizing program(s).

Congresses bring together the intellectual leadership of these programs including scientific committees, national committees for core projects of individual programs, and the staff and leaders of the central secretariats. These meetings encourage high-level dialog among the leaders of the programs, encourage development of interdisciplinary/inter-project cooperation, and help these organizations coherently implement their overall strategies.

Workshops cover a wide spectrum of activities. They may be large-scale, such as the START Young Scientists Meeting that took place prior to the ESSP OSC in 2006. They may also be smaller, focused meetings, such as the 2005 AfricanNESS Workshop, in which a pan-African group convened to begin to develop an overall framework and agenda for regional global environmental change cooperation in Africa. They also include activities focused on closely related topics such as the 2007 meeting of IPCC authors and climate experts organized by the Global Climate Observing System (GCOS), WCRP, and IGBP to discuss gaps and research needs based on analysis of the IPCC Fourth Assessment Report.

The programs assign a high priority to developing scientific capacity both in terms of young scientists and in terms of involving and fostering scientists from less-developed countries. The programs themselves, particularly through IAI and START, fund young scientist meetings and advanced training institutes all over the world. This involves bringing scientists from less-developed countries to some of the best facilities in the world and bringing top-tier global environmental change scientists to meetings and workshops throughout the developed and less-developed world. The programs, by convening many of their meetings in less-developed countries, are also developing capacity across the spectrum in those countries. While many programs are working to take advantage of the ever-increasing access to broadband communications (including e-mail, video conferencing, podcasts, social networking, etc.) the importance of in-person interactions such as those described above cannot be overemphasized.

The CCSP Interagency Working Group on International Research and Cooperation facilitates the centralized operations of and U.S. participation in the international global change research programs by serving as a channel through which "glue money" is provided to these programs. The glue money provided by CCSP and individual agencies facilitates leadership by U.S. scientists in these organizations and advances overall U.S. global change research, modeling, and observations. The U.S. funding leverages substantial funding of these programs by other countries (that in most cases is of the order of two or three times the funding provided by the United States).

The following sections describe highlights of recent activities as well as future plans of these international global change research programs and of related interagency international efforts. For more detailed information about some of these activities, see Chapter 15 of the *Strategic Plan for the U.S. Climate Change Science Program*.

HIGHLIGHTS OF RECENT ACTIVITIES

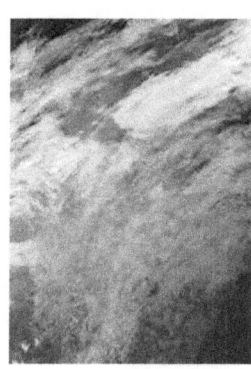

Aviation-Climate Change Research Initiative. The Aviation-Climate Change Research Initiative (ACCRI) was developed with the objective of improving the state of scientific knowledge and addressing key knowledge gaps while making the best practical use of available science and modeling capability to quantify the climate impacts of aviation. ACCRI coordinates the research efforts supported by the Federal Aviation Administration (FAA) Office of Environment and Energy and the NASA Earth Science Research Division (ESRD) as well as the NASA Applied Sciences Program (ASP), with the involvement of other Federal agencies.

Demand for air travel is projected to grow substantially. Studies based on previous passenger and cargo loads have shown that emissions from air travel as compared to other sources have been relatively small. However, concern has been raised regarding the impact of emissions from the projected increase in air traffic. To address these issues and to meet future aviation demand, the Joint Planning and Development Office of the United States developed the *Next Generation Air Transportation System (NextGen): Integrated Plan*. During June 2006, FAA and NASA jointly sponsored a workshop on the Impacts of Aviation on Climate Change to assess and document the present state of knowledge on this subject. The report documenting the recommendations and findings of this workshop is available at <web.mit.edu/aeroastro/partner/reports/climatewrksp-rpt-0806.pdf>.

ACCRI developed subject-specific white papers on key areas (covering various aspects of aviation-related climate impacts) that provided a focused, indepth review of the present understanding of the scientific principles, uncertainties, and gaps, and the present state of modeling capability based on the best current scientific knowledge of each. In 2008, a select group of scientists were convened to discuss and integrate the findings from all the white papers into a composite report on the research required to improve the state of knowledge in this area. In addition, this meeting provided suggestions for designing scenarios for impact assessment simulations and analyses using the best modeling tools and modules (and their integration). A report is expected in late 2008.

Bilateral Cooperation in Climate Change Science and Technology. Since June 2001, the United States has launched bilateral climate partnerships with 15 countries and

regional organizations that, combined with the United States, account for almost 80% of global greenhouse gas emissions. For more information on the bilateral and regional climate partnerships, visit <state.gov/g/oes/climate/c22820.htm>. Partnerships have been established with Australia, Brazil, Canada, China, Central America (Belize, Costa Rica, El Salvador, Guatemala, Honduras, Nicaragua, and Panama), the European Union, Germany, India, Italy, Japan, Mexico, New Zealand, the Republic of Korea, the Russian Federation, and South Africa. These bilateral initiatives seek to build on key elements of CCSP and the Climate Change Technology Program, including research, observations, data management and distribution, and capacity building. These partnerships now encompass 474 individual activities. Successful joint projects have been initiated in areas such as climate change science; clean and advanced energy technologies; carbon capture, storage, and sequestration; and policy approaches to reducing greenhouse gas emissions. The United States is also assisting key developing countries in efforts to build the scientific and technological capacity needed to address climate change.

Two ongoing objectives for the bilateral activities will be continued advancement of results-oriented programs and the fostering of substantive policy dialogs within all of the bilateral climate change partnerships. In order to broaden U.S. cooperative efforts to advance a practical and effective global response to climate change, the United States will expand outreach and support to the developing country community, utilizing a regional approach where feasible.

DIVERSITAS. DIVERSITAS recently established a new core project called bioGENESIS. BioGENESIS aims to develop tools for discovering, documenting, and navigating various aspects of biodiversity including elucidating the temporal and spatial aspects of how and why biodiversity evolved. The premise of the project is that study of development and evolution of biodiversity may eventually provide us with insights as to how populations and species may respond to a variety of perturbations, including global environmental change. In 2007, DIVERSITAS also conducted a series of international consultations on the development of an International Mechanism of Scientific Expertise on Biodiversity (IMoSEB). These consultations were designed to collect input from countries around the globe as to their needs and capabilities related to a global observation network for biodiversity (see <diversitas-international.org>).

Earth System Science Partnership. The ESSP—a cooperative, interdisciplinary international scientific effort (see <essp.org>)—continues to implement its four core projects on food, water, carbon, and human health in the context of global environmental change. Although the program is relatively new, scientists involved with its core projects are already producing excellent results and giving an international voice to the ESSP regarding assessments, carbon cycle science, and other areas.[1,2,3] ESSP established a Scientific Committee that is intended to help guide the activities and overall direction of the organization, and is currently developing a business plan to guide the program and aid it in the realization of its mission and goals. ESSP, its partner programs, and other global environmental change networks (APN and IAI) were invited to participate in a meeting with the Chair of UNFCCC's Subsidiary Body for Science and Technological Advice (SBSTA) and a number of country representatives to explore how SBSTA might facilitate a more effective dialog between the Parties and the research programs.

Future Climate Change Research and Observations. GCOS, with WCRP and IGBP, co-sponsored a workshop on Future Climate Change Research and Observations—held 4-6 October 2007, in Sydney, Australia—in which IPCC authors from Working Groups I and II and climate experts associated with the three sponsoring organizations participated. Based on a survey of IPCC authors' assessment of gaps, research needs, and observational needs resulting from the IPCC's Fourth Assessment Report, the workshop participants produced a set of suggestions for priority areas in research and observations that could significantly advance the science of global environmental change. The recommendations include improving and augmenting connections between global circulation models and regional models. Also recognized was the need for the climate change modeling community to improve its communication and cooperation with the community that assesses climate impacts and designs adaptation measures. Data, particularly from developing countries, were recognized as a challenge especially in terms of testing regional models. A model intercomparison project, similar to the Coupled Carbon Cycle Climate Model Intercomparison Project (C^{4}MIP), was recommended as a way to advance the science of regional climate change.

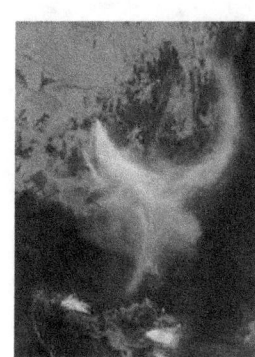

Group on Earth Observations. Ministers and officials from over 100 governments and international organizations assembled 28-30 November 2007, in Cape Town, South Africa, to advance an internationally agreed plan for building a Global Earth Observation System of Systems (GEOSS). Ministers continued to support the Group on Earth Observations (GEO) Work Plan and noted the considerable progress made since the approval of GEO in 2005 and described in the *GEO Report on Progress 2007*. They confirmed that, *inter alia*:

- The sustained operation of terrestrial, oceanic, airborne, and space-based observation networks is critical for informed decisionmaking.
- Data interoperability is critical for the improvement and expansion of observational, modeling, data assimilation, and prediction capabilities.
- Continued research and development activities and coherent planning are essential for future observation systems.

Ministers also supported establishing a process to reach consensus on the implementation of Data Sharing Principles for GEOSS and stated that the success of GEOSS depended on a commitment by all GEO partners to work together to ensure timely, global, and open access to data and products. Partners in the Integrated Global Observing Strategy (IGOS-P) reaffirmed their long-term support for the IGOS-P Themes and reconfirmed their approval of transferring the themes to GEO (see <earthobservations.org>). The international recognition, political support, and momentum enjoyed by GEO were seen as truly value-added components to the robust history of IGOS achievements.

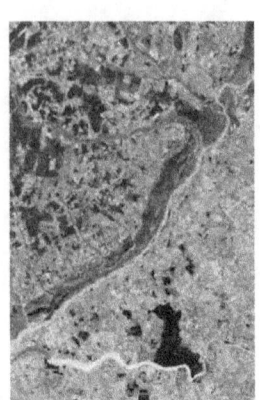

Inter-American Institute for Global Change Research. IAI was recently reviewed by an external committee of distinguished international experts convened by the American Association for the Advancement of Science (AAAS; see <aaas.org/programs/centers/sd/aaas_IAIreport_0607.pdf> for the report). The review committee found that IAI has so far produced high-quality, world-class science, and that it is effective in achieving its goals because it has become more collaborative and its projects are increasingly being led by Latin American scientists. The review committee also found that the data and information system was not realizing its goal—that is, to serve science and society and inform action. Finally, the committee found that IAI science, while increasingly interdisciplinary, did not have a strong enough emphasis on the links between environmental change and human activity. The Committee recommended that IAI maintain and, where possible, improve upon its demonstrated dedication to scientific excellence; incorporate new research around regionally relevant topics, such as risk, vulnerability, and adaptation; continue to foster regional cooperation through synthesis and other activities; develop and implement a plan to upgrade the IAI data and information system; and incorporate new research that studies "feedbacks between human activities and global and regional environmental changes."

IAI was granted funds from Canada's International Development Research Centre (IDRC) for research on land use, hydrology, and climate in the La Plata Basin, and from the MacArthur Foundation for assessment of research and institutional needs to cope with the effects of climate change on Andean biodiversity. It is also receiving increasing recognition in the region, and was invited to present findings at meetings of the region's Agriculture and Environment Ministers. See <iai.int> for more information.

International Geosphere-Biosphere Programme. IGBP (see <igbp.net>) has established a new core project called Analysis, Integration, and Modeling of the Earth System (AIMES), which evolved from the Global Analysis, Integrations, and Modeling Task Force (GAIM). AIMES, hosted by the National Center for Atmospheric Research (NCAR), focuses on quantifying the role of human perturbations of biogeochemical cycles, as well as on the overall function of biogeochemical cycles, including interactions and feedbacks with the physical climate system. IGBP, with WCRP, drafted and submitted a white paper to the IPCC regarding modeling strategies for future assessments. IGBP scientists also contributed significantly to international assessments (80 for the IPCC Fourth Assessment Report and seven co-authors of the Millennium Ecosystem Assessment).

International Human Dimensions Programme. IHDP—jointly sponsored by ICSU, the International Social Science Council (ISSC), and the United Nations University (UNU)—released its *Strategic Plan 2007-2015* (see <ihdp.org>). This new science plan emphasizes integration of the well-established ESSP core projects, cross-cutting themes, and advancing development of improved methodologies.

SyTem for Analysis, Research, and Training. START is making the transition from its first-generation effort in research-driven capacity building to the next (see <www.start.org>). The first-generation effort was based on the largely disciplinary efforts of the research programs, while the second responds to the development of ESSP and its cross-cutting projects on food, water, health, and carbon, and to the entry of new partners, such as overseas development agencies. The second-generation effort also responds to the present and growing need, demonstrated by all of the programs and broadly in governments across the globe, to more closely connect global environmental change research to society at appropriate temporal and spatial scales. By doing so, it is expected that this research may meaningfully inform decisions and policy development with the ultimate objective being to advance progress on the Millennium Development Goals.

In 2007, START continued to synthesize results from the Assessments of Impacts and Adaptation to Climate Change (AIACC) program. This effort has resulted in several synthesis books and an AIACC report to the Global Environment Facility (GEF) and the United Nations Environment Programme (UNEP). Final reports from 24 regional assessments have been or are expected to be completed and posted on START's web site in 2008. START also administered numerous capacity-building activities including sponsoring regional science planning, conducting several institutes, and organizing a young scientists meeting. The program also sponsored Ph.D. fellowships, small grants for researchers in Africa and Asia, and provided many young scientist awards.

U.S.-Japan Liaison Group on Geosciences and Environment. During 2008, the United States hosts the 12th U.S.-Japan Workshop on Global Change Research. The theme for this workshop, in Boulder, Colorado, is "Long-term Projection, Near-term Prediction, Extreme Events Projection and Observations." This activity brings together top modelers from the United States and Japan to discuss recent research accomplishments, identify gaps in knowledge, and identify unique ways in which U.S. and Japanese researchers may cooperate in order to advance the field.

World Climate Research Programme. In collaboration with IGBP and GCOS, in October 2007, WCRP convened a workshop to address the gaps and uncertainties identified in the IPCC Fourth Assessment Report in developing future observation and research requirements. The motivation for this workshop was to improve analyses of climate change risk and adaptation measures, reducing vulnerability to a changing climate. The third WCRP International Conference on Reanalysis took place in late January 2008, following an initiative by WCRP, the Japan Meteorological Agency, the Central Research Institute of Electric Power Industry, and the University of Tokyo. A reanalysis is a comprehensive global, multi-decadal data set on a regular grid generated by the latest numerical modeling and assimilation techniques to synthesize together past observations. Reanalysis data have consistent technical quality over decades and provide vital context to many types of meteorological and climatological research and applications as well as provide important insights into the usefulness and value of the climate observing system. WCRP also convened the first Workshop on Seasonal Prediction in Barcelona, Spain, in June 2007. The main objective of the workshop was to make an assessment of current skill in seasonal prediction, with particular emphasis on surface temperature and precipitation. A WCRP Workshop on Global Prediction of the Cryosphere took place in October 2007, to identify gaps in current understanding of how sea ice, ice shelves, glaciers, snow cover, lake ice, river ice, permafrost, and the large ice sheets of Greenland and Antarctica might change in future. These—and the four WCRP core projects CLIVAR, GEWEX, CLiC, and SPARC—are a few of the many WCRP international activities in support of climate research. See <wcrp.wmo.int> for more information.

HIGHLIGHTS OF PLANS FOR FY 2009

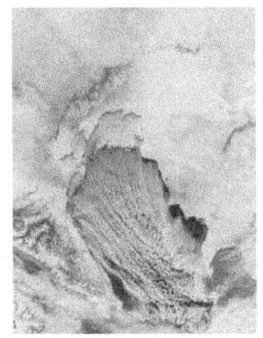

Arctic Observing Network. The Arctic is experiencing unprecedented system-wide change; change that has few equals elsewhere on Earth. This change has global implications and continued changes will have significant regional and global environmental and societal consequences. The dramatic recession of Arctic sea ice cover that took place in summer 2007 is one example illustrating the enormous scale of recent Arctic change, but the *Arctic Climate Impact Assessment*, released in 2005, highlights potential impacts of changes in the sea ice cover as well as many other environmental parameters (see <www.eol.ucar.edu/projects/aon-cadis> and <www.acia.uaf.edu>). Monitoring polar climate and understanding its feedbacks are key priorities described in the CCSP Strategic Plan.

A recent National Research Council (NRC) report, *Towards an Integrated Arctic Observing Network*, concluded that current Arctic observing systems are not capable of characterizing the change that is now in motion and they do not provide the data necessary to enable scientific synthesis and modeling studies that are essential for better understanding the regional and global causes and consequences of Arctic change. CCSP supports creation of a comprehensive Arctic Observing Network (AON). NSF along with the twelve other Federal agencies that make up the Interagency Arctic Research Policy Committee (IARPC) are engaged in a wide range of Arctic-observing activities. Together they are implementing an interagency activity entitled the Study of Environmental Arctic Change (SEARCH; see <arcus.org/search>) to better understand climate change as identified in the *Arctic Climate Impact Assessment*. AON is also being put forth as a U.S. contribution to a multi-nation, pan-Arctic observing network. It will represent a lasting legacy of the International Polar Year (IPY) and will contribute to broader international goals surrounding the establishment of GEOSS and the Global Ocean Observation System (GOOS).

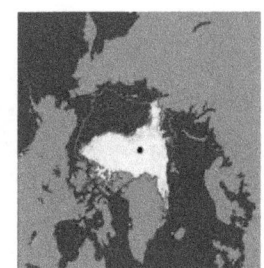

Aviation-Climate Change Research Initiative. Based on the 2008 composite report, model simulations and analyses will be performed to quantify climate impacts of aviation. A multi-model and multi-team approach will be adopted to support this activity. Long-term research activities will be implemented as needed during 2009 and beyond. These research activities will address key information gaps and make practical use of improved scientific knowledge and modeling capability to help characterize and mitigate aviation's climate impact.

African Network on Earth System Science. Since its inception in 2005, AfricanNESS has developed a science plan that will serve as a roadmap for sustained regional global environmental change research in Africa. AfricanNESS, with the ICSU Regional Office for Africa and facilitated by IGBP, released a merged science plan in 2008. The merging

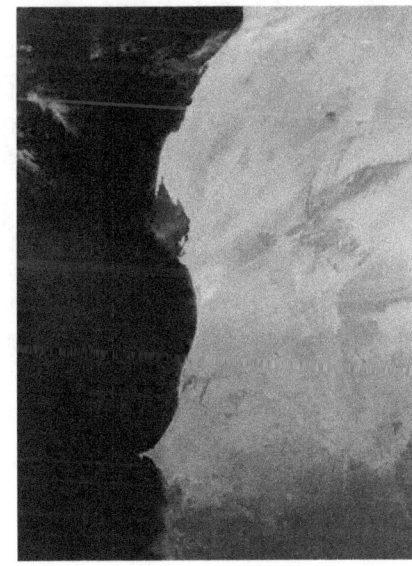

of these science plans is the result of several years of parallel effort and significant community participation. It is hoped that the merged science plan will be widely accepted and implemented. Many challenges remain, but in those challenges lie significant opportunities to advance science globally. One is finding ways to rebalance and develop research capacity. The science plan focused on four cross-cutting areas: food and nutritional security, water resources, health, and ecosystem integrity. A significant number of global environmental change researchers currently in Africa could contribute to AfricaNESS projects.

Atlantic Interoperability Initiative for Reducing Emissions. The Atlantic Interoperability Initiative for Reducing Emissions (AIRE) partnership is a collaboration of the European Commission, FAA, airlines, and aviation industry partners working together to speed development and implementation of environmentally friendly new technologies and operational procedures that reduce engine exhaust emissions and associated noise. Approximately 2 to 3% of overall global greenhouse gas emissions can be attributed to aviation; expected traffic growth may increase this contribution. To counter this effect, AIRE will demonstrate maturing air traffic control infrastructure technologies that will reduce release of aircraft engine exhaust emissions through improved system efficiency and/or operations. These technologies are being demonstrated for each unique segment of flight operations: ground/surface taxi movements, oceanic (en route) cruise, and arrival landing operations. The overall objective is to enhance surface movement operational efficiency, save fuel, and reduce engine exhaust emissions and associated noise for international flight operations using the system advances identified. Collaborative demonstration flights should begin in FY 2009.

DIVERSITAS. DIVERSITAS will continue to advance international biodiversity science through cooperation with U.S. agencies and international organizations, and contributions to international conventions. DIVERSITAS expects to continue to work with NASA as the lead for the biodiversity task of GEOSS that involves development of a science plan and implementation strategy for a global biodiversity observing system. It is expected that once this plan is completed near the end of 2008, attention will be turned to implementation of the observing system. DIVERSITAS will continue its collaboration with the Convention on Biological Diversity (CBD). In the margins of the CBD 9th Conference of the Parties (COP9). DIVERSITAS, together with the International Union of Biological Science (IUBS) and the German government, organized a 3-day scientific conference, leading to a formal declaration to COP9 delegates (Bonn, May 2008). The second DIVERSITAS Open Science Conference is scheduled for 13 to 16 October 2009, in Cape Town, South Africa.

Earth System Science Partnership. A review of ESSP was recently commissioned by ICSU and IGFA at the request of its partner programs. Given that the program is still relatively new and has undergone substantive recent changes, the review focuses on providing strategic advice as to options for its future development. The report will be released in 2008. ESSP expects to continue to implement its core projects on carbon, waters, health, and food; continue implementation of the Monsoon Asia Regional Study; and to develop and release a business plan. The Scientific Committee will also have an increasing role in development of the overall program, including realization of its mission and goals.

ESSP agreed to convene a regular series of seminars at each SBSTA meeting. The topics of the seminars would be developed collaboratively between ESSP, SBSTA, and the UNFCCC Parties. The seminars would be followed by a discussion session in which the Parties and the research community would have an opportunity for dialog. All agreed that IPCC should remain the primary assessment mechanism for UNFCCC, but these seminars would be an effective way to facilitate regular information exchange between the Parties and the research community. In addition to the meeting, ESSP (spearheaded by WCRP) organized a side event entitled "Connecting Earth System Science Research to Climate Change Policy." The side event featured four seminars on different aspects of climate change research and connections to society.

Inter-American Institute for Global Change Research. The Minister of the Environment of the Dominican Republic will host an IAI Strategic Planning session in 2008 to chart out the Institute's next decade, and preliminary drafting sessions are underway with input from the Member Countries and the IAI's Scientific Advisory Committee. Together with APN, IAI will meet with the Chair of SBSTA to facilitate dialog between the international research programs and the UNFCCC Parties. IAI, together with IGBP and ICSU, are also consulting with the Amazon Treaty Organization (OTCA) on a similar issue for the Amazon Basin.

IAI is in the midst of planning a second IAI-NCAR Colloquium for the second half of 2008 on "Seasonality and Water Resources in the Western Hemisphere," to be held in Mendoza, Argentina. The IAI's successful Training Institutes series continue with intensive sessions on Information Management (Panama, February 2008), Adaptation Risks (side event to Central American Presidential Summit, Honduras, May 2008), Semi-Arid Water Management (Brazil, October 2008), and Urban Responses to Climate Change (Chile, spring 2009).

Discussions with the United Kingdom's Department for International Development (DFID) will continue on water and health projects, and may be linked with the IAI-IDRC project to consolidate findings from several IAI research networks and develop new plans linked to national priorities for science and capacity building for climate change and adaptation.

IAI will continue discussions with the IGFA agencies on strategies to stably fund relevant international global change programs and avoid 'double-dipping' when these programs apply for funds from both the IGFA agencies and the regional institutes funded by the same agencies.

International Geosphere-Biosphere Programme. IGBP held its Fourth Congress, "Sustainable Livelihoods in a Changing Earth System," in Cape Town, South Africa, 5-9 May 2008. Anticipated participants include the Steering Committee, the Scientific Steering Committees, National Committees, partners, and African stakeholders. The meeting had many objectives but was primarily aimed at improving linkages, communication, and integration across all IGBP efforts. The meeting also explored the connections between environmental change and development on a regional basis in Africa.

It is expected that a review of the IGBP, concurrent with that of the WCRP, co-sponsored by ICSU and IGFA, will begin in the FY 2009 time frame.

International Human Dimensions Programme. IHDP is currently undergoing a synthesis effort for its Industrial Transformation (IT) and Global Environmental Change and Human Security (GECHS) projects. The program has applied new cross-cutting themes including vulnerability, resilience and adaptation, governments and institutions, social learning and knowledge, and thresholds and transitions. At the same time, IHDP has also applied a new focus on methodologies that will cut across all of its core projects. The program will focus on enhancing statistical methods, improving simulations, incorporating case studies and narratives, and applying systems analysis, as well as configuration and comparative analysis. Overall, IHDP will emphasize more involvement with international assessments including IPCC and enhancing linkages with major social science research themes and communities.

Northern Eurasia Earth Science Partnership Initiative. The Northern Eurasia Earth Science Partnership Initiative (NEESPI) is an External Project of the IGBP (see <neespi.org>). It is a multidisciplinary program of internationally supported Earth systems science research focusing on issues in northern Eurasia relevant to regional and global scientific and decisionmaking communities. Northern Eurasia is undergoing significant changes associated with a rapidly warming climate in this region and with important changes in governmental structures since the early 1990s and their associated influences on land use and the environment across this broad expanse of the Earth. The NEESPI research strategy intends to capitalize on a variety of remote-sensing and other tools. NEESPI will implement a general modeling framework that links socioeconomic factors with

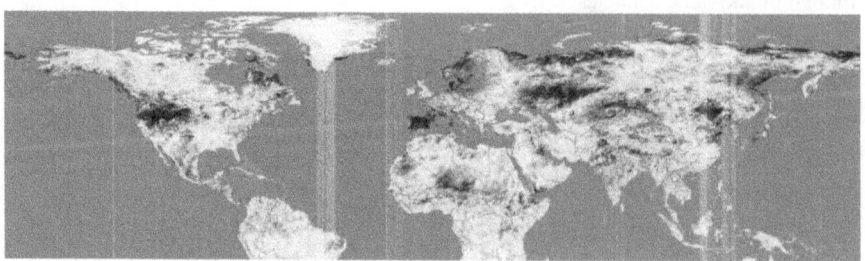

models such as crop, pollution, land use, ecosystem, and climate with observational
data to address the key research questions within northern Eurasia:

- How will future human actions affect the Northern Eurasia and global ecosystems?
- What will be the consequences of global changes for the regional environment, the
economy, and the quality of life in northern Eurasia?

Currently, over 100 NEESPI projects involve about 400 investigators from 30 countries,
including the United States, the Russian Federation, the People's Republic of China,
the European Union, Japan, and Canada. NEESPI scientists are active within several
IPY activities on cold land processes studies, terrestrial hydrology, atmospheric
aerosols, and biospheric processes. NEESPI has also established several Focus Research
Centers on specific themes and Regional Focus Research Centers in northern Eurasia.

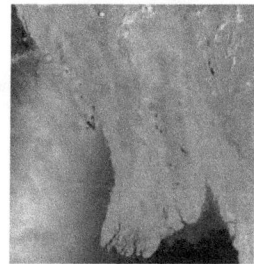

Monsoon Asia Integrated Regional Study. The Monsoon Asia Integrated Regional Study
(MAIRS) is an Integrated Regional Study project under ESSP that includes IGBP, IHDP,
WCRP, and DIVERSITAS components. The region of monsoon Asia covers South,
Southeast, and East Asia. With the highest population density of any comparable region
of the world, this region has experienced one of the most rapid environmental changes
in the past decade and is likely to undergo further rapid economic development in the
coming years. Human activities in the monsoon Asia region significantly affect
environmental conditions, both regionally and globally.

The goal of MAIRS is to better understand how human activities in the region are
interacting with and altering the natural variability of atmospheric, terrestrial, and
marine components of the monsoon system; to contribute to the provision of a sound
scientific basis for sustainable development in monsoon Asia; and to develop a predictive
capability for estimating changes in global-regional linkages in the Earth system and to
project the consequences of such changes.

NASA has provided support for the MAIRS program by soliciting proposals studying
the regions within the MAIRS domain and by supporting MAIRS logistics. A NASA-
MAIRS joint meeting is planned for the fall of 2008.

Polar Earth Observing Network. The Polar Earth Observing Network (POLENET) will
provide ground-based seismic and Global Positioning System (GPS) observing networks
in Antarctica and a GPS network in Greenland (see <polenet.org>). POLENET is
designed to directly measure solid earth phenomena needed to eliminate sources of
uncertainty in satellite-derived measurements and models. The observing network is

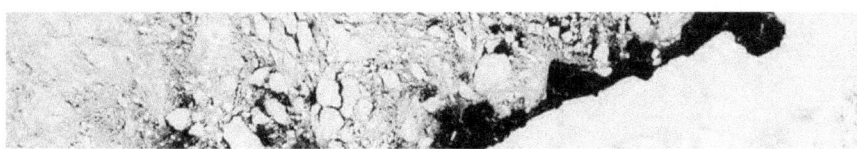

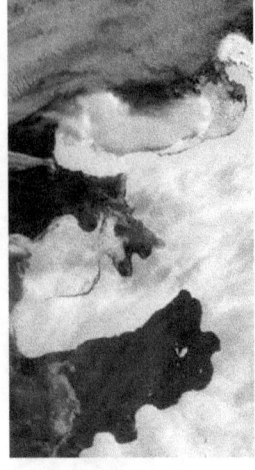

intended to serve as a lasting legacy of IPY and will contribute significantly to ongoing assessments of climate change. It also contributes to broader international goals surrounding the establishment of GEOSS.

SysTem for Analysis, Research, and Training. In 2008 and beyond, START will continue its collaboration with its co-sponsors on its Advancing Capacity to Support Climate Change Adaptation (ACCCA) program. Fourteen projects have been funded and are underway. Five additional projects are under review. START will also continue its new African Climate Change Fellowship program, sponsored by the IDRC's Climate Change Adaptation in Africa program. START is also developing many collaborations with organizations such as the Stockholm Environment Institute, APN and IAI, IHDP, and IGBP at its upcoming Congress in South Africa.

World Climate Research Programme. WCRP, as decided by the Joint Scientific Committee, has as its highest priority the implementation of its cross-cutting activities. These activities include study of monsoons, anthropogenic climate change, atmospheric chemistry and climate, sea-level rise, decadal climate predictability, and IPY. WCRP has also welcomed a new director who started in the beginning of 2008. It is expected that a review of the program, co-sponsored by ICSU and IGFA, will begin in the FY 2009 time frame.

INTERNATIONAL RESEARCH AND COOPERATION CHAPTER REFERENCES

1) **Raes**, F. and R. Swart, 2007: Climate assessment: what's next? *Science*, **318(5855)**, 1386, doi:10.1126/science.1147873.

2) **Piao**, S., P. Ciais, P. Friedlingstein, P. Peylin, M. Reichstein, S. Luyssaert, H. Margolis, J. Fang, A. Barr, A. Chen, A. Grelle, D.Y. Hollinger, T. Laurila, A. Lindroth, A.D. Richardson, and T. Vesala, 2008: Net carbon dioxide losses of northern ecosystems in response to autumn warming. *Nature*, **451**, 49-53, doi:10.1038/Nature06444.

3) **Canadell**, J.G., C. Le Quéré, M.R. Raupach, C.B. Field, E.T. Buitenhuis, P. Ciais, T.J. Conway, N.P. Gillett, R.A. Houghton, and G. Marland, 2007: Contributions to accelerating atmospheric CO_2 growth from economic activity, carbon intensity, and efficiency of natural sinks. *Proceedings of the National Academy of Sciences*, **104(47)**, 18866-18870.

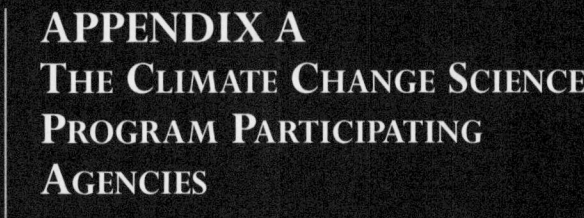

APPENDIX A
THE CLIMATE CHANGE SCIENCE PROGRAM PARTICIPATING AGENCIES

APPENDIX A

THE CLIMATE CHANGE SCIENCE PROGRAM PARTICIPATING AGENCIES

The following pages present information about the contributions to the CCSP of each of the program's participating agencies:

- Department of Agriculture (USDA)
- Department of Commerce (DOC)
- Department of Defense (DOD)
- Department of Energy (DOE)
- Department of Health and Human Services (HHS)
- Department of the Interior (DOI)
- Department of State (DOS)
- Department of Transportation (DOT)
- Agency for International Development (USAID)
- Environmental Protection Agency (EPA)
- National Aeronautics and Space Administration (NASA)
- National Science Foundation (NSF)
- Smithsonian Institution (SI).

Principal Areas of Focus, Program Highlights for FY 2009, and Related Research are summarized for each agency.

U.S. DEPARTMENT OF AGRICULTURE

Agricultural Research Service (ARS)
Cooperative State Research, Education, and Extension Service (CSREES)
Economic Research Service (ERS)
Forest Service (FS)

Principal Areas of Focus

The U.S. Department of Agriculture (USDA) conducts and sponsors a broad range of research that supports the Climate Change Science Program (CCSP). USDA research focuses on questions that are relevant to decisionmakers at the Federal, State, and local levels. Areas of emphasis include evaluating risks to natural resources, estimating the role of forestry and agricultural activities in greenhouse gas emissions and carbon sequestration, and developing practical management strategies and approaches to manage emissions and adapt to changes. USDA's research program seeks to determine the significance of terrestrial systems in the global carbon cycle; promotes the capture and use of methane emitted from livestock waste facilities for on-farm power generation; assesses the potential of bioenergy as a substitute for fossil fuels; identifies agricultural and forestry activities that can help reduce greenhouse gas concentrations and increase carbon sequestration; quantifies the risks and benefits arising from environmental changes to agricultural lands and forests; and develops management practices that can adapt to the effects of global change, including potential beneficial and adverse effects.

Program Highlights for FY 2009

ARS's climate change research focuses on the carbon cycle and carbon storage; trace gas emissions and sinks; impacts of change on agricultural systems; and feedbacks among agriculture, weather, and the water cycle. During FY 2008, ARS began formulating its next 5-year national plan for global change research via a stakeholder-scientist workshop, creation of a national program action plan, and development of individual research projects. ARS research continues to build a scientific knowledge base for decision-support technologies. ARS global change research enables producers, land managers, and strategic decisionmakers to successfully mitigate the contributions of agricultural systems to climate change and successfully adapt agricultural systems to its potential effects. The **G**reenhouse gas **R**eduction through **A**gricultural **C**arbon **E**nhancement **net**work (GRACEnet) project is being conducted at 30 locations across the United States, measuring greenhouse gas emissions from multiple agricultural management systems, and allowing the formulation of guidelines for mitigating emissions and optimizing carbon sequestration. The impact of elevated atmospheric carbon dioxide on agroecosystems, such as increased pressure from weeds and invasive species, will continue as a research component. Responses of the hydrologic cycle to climate change that may affect soil water availability for agriculture and other water supplies, such as drought, will be investigated. Development of environmentally friendly and economically feasible alternatives to stratospheric ozone-depleting methyl bromide as a treatment to control pests will continue. ARS participates in intragency and interagency working groups to ensure relevant and significant contributions to the understanding, response to, and mitigation of global change and its impact on production of food, fiber, bioenergy, and natural resources.

The CSREES Program on Global Change and Climate integrates research, education, and extension expertise with new approaches that are economically sound and environmentally advantageous. CSREES conducts its programs primarily in partnership with land-grant university scientists and cooperative extension faculty. Programs such as UV-B Monitoring and Research strengthen the Nation's capacity to address critical environmental priorities and contribute to improved air, soil, and water quality; fish and wildlife management; enhanced aquatic and other ecosystems; the sustainable use and management of forests, rangelands, watersheds, and other renewable natural resources; and a better understanding of global climate change, including its impact on the diversity of plant and animal life. CSREES efforts also demonstrate the benefits and opportunities of sustainable development, and contribute to the economic viability of agriculture and rural communities realizing the impact of environmental policies and regulations. The sustainability of agriculture, forest, and rangelands depends on understanding the factors that influence climate change, the mechanisms that may enhance or mitigate this change, and its effects on food and fiber production and natural resources. Program priorities are drawn primarily from the *Strategic Plan of the U.S. Climate Change Science Program* and include human dimensions. Competitive funding for Global and Climate Change Research is provided through the National Research Initiative (NRI) Competitive Grants Program, which is charged with funding research, education, and extension activities to address key problems of national and regional importance in biological, environmental, physical, and social sciences relevant to agriculture, food, the environment, and communities on a peer-reviewed, competitive basis. To address these problems, NRI advances scientific knowledge in support of agriculture, forestry, and related topics. The program also supports the integration of research, education, and extension to generate, translate, and transfer new technology and knowledge into practical applications focused on solving problems of national importance. Past Global and Climate Change initiatives include carbon cycle science, land-use and land-cover change, and invasive species.

Forest Service research is concentrated on three areas. First, mitigation research aims to increase the fossil-fuel carbon removed from the atmosphere by forests and by offsets to fossil fuels provided by forest products. Second, adaptation research aims to reduce emissions of forest carbon from major disturbances by developing and evaluating methods to increase ecosystem resilience to current and future climate stresses on forests and rangelands, also thereby maintaining ecosystem health and services (e.g., timber, water supplies, biodiversity). Third, creation of decision-support systems—including monitoring, reporting, and synthesis of information—supports land managers and policymakers in adopting these new research results for optimum management of forests and rangelands under a changing environment. Within these three areas, Forest Service research works at (i) expanding understanding of the global carbon cycle in forest and rangeland ecosystems, and the consequences and feedback from the management and use of these ecosystems as they interact with the atmosphere; (ii) improving accuracy and ease of analyses of U.S. forest carbon inventory, and other monitoring and analysis systems for carbon dioxide; (iii) enhancing understanding of climate change impacts on forest health, major disturbance regimes, and ecosystem services; (iv) integrating observation and monitoring networks with process studies to better understand, forecast, and manage relationships between forest and rangelands and climate; (v) accelerating the development of management technologies to increase carbon sequestration, provide fossil-fuel offsets, enhance forest productivity, and maintain environmental quality; and (vi) providing integrated prediction models of forest dynamics under expected future changes in climate and atmospheric chemistry.

Related Research

USDA remains active in the Climate Change Technology Program (CCTP) and related efforts. The Forest Service, Natural Resources Conservation Service (NRCS), ARS, CSREES, and the Rural Development mission area support improved measurement and accounting of greenhouse gases from agriculture and forestry systems, as well as energy initiatives and renewable energy systems such as biofuels and biomass-related research and development. NRCS and the Forest Service are cooperating in development of web-based assessment tools for agricultural producers to account for benefits accruing on carbon fluxes and greenhouse gas emission from conservation practices. In addition, NRCS and the Forest Service are developing new measurement technologies, analytical techniques, and information management systems related to spatial carbon distributions. USDA also is filling gaps in ecosystem information by continuing to collect data on land use, resource conditions, and climate through the National Resources Inventory, the Forest Inventory and Analysis Program, the Soil Climate Analysis Network, and the Snowpack Telemetry system. These networks provide critical data needs on the status and condition of land use in the United States in support of CCSP research.

DEPARTMENT OF COMMERCE

Principal Areas of Focus

The National Oceanic and Atmospheric Administration (NOAA) and the National Institute of Standards and Technology (NIST) comprise the Department of Commerce contribution to the CCSP.

NOAA's climate mission is to "understand climate variability and change to enhance society's ability to plan and respond." This is an end-to-end endeavor whose overall objective is to provide decisionmakers with a predictive understanding of the climate and to communicate climate information so that the public can incorporate it into their decisions. These outcomes are achieved through implementation of a global observing system, focused research to understand key climate processes, improved modeling capabilities, and the development and delivery of climate information services. NOAA aims to achieve its climate mission and outcomes through the following objectives:

- Describe and understand the state of the climate system through integrated observations, monitoring, and data management
- Understand and predict climate variability and change from weeks to decades to a century
- Improve the ability of society to plan for and respond to climate variability and change.

NOAA relies on its Federal, academic, private, and international partners to achieve its objectives. These objectives are implemented through three distinct, yet integrated, programs: Climate Observation and Monitoring, Climate Research and Modeling, and Climate Services Development.

NIST provides measurements and standards that support accurate and reliable climate observations. NIST also performs calibrations and special tests of a wide range of instruments and techniques for accurate measurements. In FY 2009, NIST is included as a discrete element of the CCSP cross-cut to provide specific measurements and standards of direct relevance to the program.

Program Highlights for FY 2009

National Oceanic and Atmospheric Administration

Climate Observation and Monitoring

The Climate Observation and Monitoring Program develops and sustains integrated atmospheric, oceanic, and Arctic observation networks, primarily *in situ*, and maintains consistent and long-term archive and access to historical climate data. Examples of NOAA observation networks include the U.S. Climate Reference Network and the carbon dioxide (CO_2) baseline observatories, including Mauna Loa and South Pole stations. NOAA routinely provides climatological information, such as basic statistics and extremes, based upon extended records usually greater than 30 years in length. The program has two basic capabilities: observations (atmosphere, oceans, and forcing), and data management and information. These capabilities taken together increase the value and utility of both *in situ* and satellite observations, improve the performance of models, and reduce the uncertainty of predictions. The program contributes to the national and global objectives outlined in the *Strategic Plan for the Climate Change Science Program*, as well as NOAA's Strategic Plan, the *Strategic Plan for the U.S. Integrated Earth Observation System (IEOS)*, and the *Global Earth Observation System of Systems (GEOSS) 10-Year Implementation Plan*.

Activities in FY 2009 will:

- Create a scientific data stewardship plan to generate, analyze, and archive data from climate satellite sensors in long-term climate data records
- Maintain the Global Ocean Observing System (GOOS)
  - Sustain progress toward completing the U.S. contribution to the international GOOS
  - Continue technology refresh to replace obsolete components of the Tropical Atmosphere Ocean Array, a critical El Niño-Southern Oscillation climate observing system
- Maintain the U.S. Climate Reference Network and continue the installation of soil moisture sensors at 114 stations in support of the National Integrated Drought Information System
- Continue to re-measure key ocean properties along cross-sections in the South Atlantic and North Pacific that were last measured in 1989 and 1991, respectively, via the Repeat Hydrography Program
- Continue to maintain and update Carbon Tracker, the combined measurement and modeling system that keeps track of the emissions ("sources") and removal ("sinks") of atmospheric CO_2 globally
- Integrate the North American Carbon Program and relevant aspects of the Ocean Carbon and Climate Change Program to better quantify and understand the carbon budget of North America and adjacent ocean basins, including terrestrial, freshwater, oceanic, and atmospheric sources and sinks that influence atmospheric CO_2 and methane (CH_4).

Climate Research and Modeling

The Climate Research and Modeling Program assimilates observations to produce retrospective and current analyses of climate conditions, examines the attribution of climate events, and develops models to make climate predictions and projections relevant to users. The program draws upon three capabilities: understanding climate processes; Earth system modeling, predictions, and projections; and analysis and attribution. The program maintains a suite of operational climate outlooks and strives to implement the next-generation operational climate outlooks and assessments by improving climate models, improving forecast generation techniques, and maintaining real-time climate monitoring data sets. Two essential components of the program are the global climate analysis data sets generated by synthesizing diverse data sources using state-of-the-art forecast models, and the regular and systematic attribution of causes of past, current, and evolving climate conditions using modern climate diagnostic techniques.

Activities under this program leverage an extensive array of peer-reviewed, university-based competitive research activities to understand atmospheric and oceanic processes, both natural and human-related. Research may be directly applied to climate projection and to policy decisions, and provides timely and adequate information.

This program provides the Nation with a seamless suite of environmental forecasts and projections from intraseasonal to multidecadal time scales and regional to global spatial scales. The program helps regional and national managers to better plan for the impacts of climate variability, and to provide climate assessments and projections to support policy decisions with objective and accurate climate change information.

Activities in FY 2009 will:

- Continue model experiments and paleoclimate research on potential mechanisms, patterns, causes, and impacts of abrupt climate change events
- Support research to assess the Atlantic Meridional Overturning Circulation variability and its implications for rapid climate change

- Continue to construct high-quality reanalysis of the coupled ocean-atmosphere system from the start of the satellite era (late 1970s) through 2007
- Provide global climate analyses required to describe major features of 20th century climate and the capacity to address the causes of observed regional climate variations under the project Explaining Climate to Improve Predictions (ECIP)
- Study the interactions between aerosols and non-CO_2 gases, enhanced measurements of atmospheric water vapor, and interactions of pollutants with climate change
- Continue to focus on the calibration and validation of research-mode ensemble forecasting techniques for surface and subsurface hydrological parameters, especially on longer seasonal time scales
- Complete a coupled ocean/sea-ice model based on new ocean model code base
- Complete a high spectral resolution radiation model for use in next-generation attribution
- Continue to understand the transport and properties of absorbing aerosols and their precursors to the Arctic polar region in an effort to quantify the contribution of absorbing aerosols to the melting of Arctic ice
- Continue analysis of isotopes and other tracers to quantify the uptake of anthropogenic CO_2 by the global ocean and its distribution within the ocean's interior
- Improve the representation of two key processes in climate models in order to predict the future magnitudes of the two largest global carbon sinks: fertilization of forests by increasing atmospheric CO_2, and draw-down of carbon in the Southern Ocean resulting in its storage in intermediate and deep waters.

Climate Services Development

The Climate Services Development Program assesses impacts of climate variability and change, supports regional adaptation strategies, and develops climate information products and tools appropriate for evolving user needs. The program supports decisionmakers by providing information to improve management of the sectors and geographic areas that are sensitive to impacts from weather and climate. Management issues include annual losses from droughts and floods, heat and cold waves, the positive and negative impacts of El Niño and La Niña events, sea-level rise, and other high-impact climate events. The information the program provides includes assessments, supporting data sets, and stakeholder-driven research and applications.

The program addresses an increased demand for traditional climate services such as data and forecast dissemination and customer support, as well as identifying and satisfying new requirements for information on long-term climate trends; linkages between climate variability, climate change, and weather extremes; assessments of vulnerability; and decision support in sectors such as drought and water management, fire, emergency preparedness, health, transportation, energy, coastal, urban, and ecosystem management. Demand for increased services is met through research of decisionmaker needs, prototype product development, transition of research products into operations, and operational delivery and support. The National Integrated Drought Information System (<www.drought.gov>) is a new example of NOAA's leadership in the development of issue-focused climate services.

The program links producers and users of climate information, allowing decisionmaker-inspired creation of new knowledge, processes, tools, and products to improve planning, risk management, resource allocation, impacts assessment, adaptation, mitigation, early warning, and operational response in sectors sensitive to climate variability and change. The program relies heavily on NOAA's extensive U.S. infrastructure with more than 150 offices at the national, regional, and local levels contributing.

The program also leverages partners at the international, national, regional, State, and local levels, as well as academia.

Activities in FY 2009 will:
- Develop decision-support tools, such as climate-fisheries models and web-based climate-agriculture tools
- Increase the understanding of regional climate impacts through basic and applied research and diagnostic and modeling studies
- Develop methodologies and decision-support tools focused on sea-level rise, extreme events and community planning, climate and integrated ecosystem management, climate extension in coastal regions, and climate and urban drainage systems
- Continue—through the Regional Integrated Sciences and Assessments Program—to enhance integration of regional programs with applied research capabilities to strengthen end-to-end development and provision of climate information services
- Continue to develop the National Integrated Drought Information System
  - Continue development of the U.S. Drought Portal in conjunction with Federal partners (USGS, NASA, and USDA) and non-Federal partners (Drought Mitigation Center and the Earth Systems Integrated Enterprise)
  - Continue development of the next-generation Climate Forecast System (CFS) and of a multi-model ensemble prediction system with CFS to accelerate the transition of research advances to new and improved objective drought monitoring and prediction products.

National Institute of Standards and Technology

NIST provides measurements and standards that support accurate and reliable climate observations. In FY 2009, NIST will address critical gaps in climate change science that are limiting long-term climate policy decisionmaking by:
- Resolving discrepancies in satellite measurements of radiation, including solar irradiance, reflected solar radiation, outgoing longwave radiation, and surface radiation
- Providing critical information about aerosols and atmospheric components believed to play a major role in global climate change.

DEPARTMENT OF DEFENSE

Principal Areas of Focus

The Department of Defense (DOD)—while not supporting dedicated global change research—continues a history of participation in the CCSP through sponsored research that concurrently satisfies national security requirements and stated goals of the CCSP. All data and research results are routinely made available to the civil science community. DOD science and technology investments are coordinated and reviewed through the Defense Reliance process and published annually in the *Defense Science and Technology Strategy*, the *Basic Research Plan*, the *Defense Technology Area Research Plan*, and the *Joint Warfighting Science and Technology Plan*.

Program Highlights for FY 2009

Satellite Sensors and Observations

Via the Air Force, DOD funds 50% of the National Polar-Orbiting Operational Environmental Satellite System (NPOESS)—a result of the convergence of national sensing suites. NPOESS will monitor global environmental conditions, and collect and disseminate data related to weather, atmosphere, oceans, land, and near-space environment. The NPOESS Program is managed by the tri-agency Integrated Program Office run by DOC, DOD, and NASA.

Global Observations and Models

The Navy is a principal member of the National Oceanographic Partnership Program, incorporating the Integrated Ocean Observing System and associated data management and communications, the Global Ocean Observation System, the Global Ocean Data Assimilation Experiment (GODAE), and the National Federation of Regional Associations (<www.ocean.us> and <usnfra.org>). This broad partnership of agencies collaborates in the development and demonstration of integrated ocean observations systems, data management systems, and real-time coastal, basin-scale, and global ocean prediction systems. As part of GODAE, the Navy funds development of the Hybrid Coordinate Ocean Model (HYCOM), a predictive model which runs efficiently in parallel computing environments and includes sophisticated techniques for assimilation of satellite and *in situ* observations. The U.S. GODAE data server (see <usgodae.org>) has been funded by the Navy for the duration of GODAE activities, which come to a close in 2008. The goal of the data server project is to develop and implement a comprehensive data management and distribution strategy that allows easy and efficient access to HYCOM-based ocean prediction system outputs to coastal and regional modeling sites, making them available to the wider oceanographic and scientific community, including climate and ecosystem researchers, and the general public, especially students in middle and high schools.

The Office of Naval Research (ONR) supports a number of research programs that, while directed toward fulfilling the objectives outlined in the *Naval Science and Technology Strategic Plan* (available at <www.onr.navy.mil>), also project onto the science goals of the CCSP. Within the "Operational Environments" focus area, ONR incorporates both observational and modeling elements into major field programs designed to learn more about the underlying physics of the ocean and atmosphere. The development of new sensors, sensing platforms, and sensing strategies is supported in order to achieve these goals, and ongoing research into predictive systems for the ocean and atmosphere is supported, with the primary goal of improving environmental forecasts for DOD. Most of this basic research enhances fundamental understanding of the dynamics of the coupled ocean-atmosphere system, and is thus relevant to climate change issues. A recent outcome of the program is a new coupled ocean-wave-atmosphere model for hurricanes which shows significant promise in improving forecasts of storm intensity. This work is being continued in FY 2009 under two new 5-year initiatives to study the generation of tropical cyclones and their impact on the thermal structure of the upper ocean. The research may lead to a better representation of these systems in climate simulations, and improved understanding of the sensitivity of these high-impact weather events to subtle changes in the Earth's climate.

Polar Regions Research

The *Strategic Plan for the Climate Change Science Program* highlights polar and sub-polar regions as research emphases, since they have exhibited more rapid changes than the lower latitudes. The U.S. Army Cold Regions Research and Engineering Laboratory (CRREL) is America's lead Federal laboratory for polar and sub-polar research. The CRREL research program responds to the needs of the military, but much of the research also benefits the civilian sector and is funded by non-military customers such as NSF, NOAA, NASA, DOE, and State governments. DOD research has examined impacts of climate change on retreating Arctic sea ice. Satellite data show that the extent of Arctic sea ice has decreased by about 10%, and the sonar data collected by U.S. Navy submarines in the Arctic between 1957 and 2000 show the average ice thickness has decreased between 33 and 42%. CRREL and the University of Alaska are developing a web-accessible Alaska Engineering Design Information System—an analytic toolkit for engineers that presents a broad array of geospatial terrestrial, oceanic, and atmospheric environmental data in a geographic information system.

The Navy, through its participation in the National Oceanographic Partnership Program, funds research related to climate change under the call for proposals "Coastal Effects of a Diminished-Ice Arctic Ocean." The funded efforts explore ocean observing system strategies for the Alaska Beaufort and Chukchi Seas, changes in the circulation and wave dynamics of the coastal Arctic, the impact on coastal production and sediment transport, and the measurement and prediction of seasonal changes in sea-ice cover in the Beaufort and Chukchi Seas.

Related Research

Other DOD-sponsored research and supporting infrastructure also contribute to observing, understanding, and predicting environmental processes related to global change. Associated research programs include theoretical studies and observations of solar phenomena, monitoring and modeling of unique features in the middle and upper atmosphere, terrestrial and marine environmental quality research, and energy conservation measures. DOD's continued investment in environmental infrastructure—such as the Oceanographic Research Vessel Fleet, and the various services' operational oceanographic and meteorological computational centers—will continue to provide data and services useful to CCSP.

DEPARTMENT OF ENERGY

Principal Areas of Focus

Research supported by the Department of Energy (DOE) Office of Science focuses on the effects of energy production and use on the global climate system. The research seeks to understand the regional and global climate response to changes in greenhouse gas and aerosol concentrations. Research covers three program areas: 1) climate change modeling, 2) climate forcing, and 3) climate change response.

Program Highlights for FY 2009

DOE will enhance and continue support of climate change research at its National Laboratories and other public and private research institutions, including universities. In support of CCSP, the DOE Office of Science's Climate Change Research Program will continue to provide the data and predictive understanding that will enable objective, scientifically rigorous assessments of the effects on climate of human-induced forcing due especially to energy-related emissions and the potential consequences of human-induced climate change.

Climate Change Modeling

DOE will enhance climate modeling research to develop, improve, evaluate, and apply fully coupled atmosphere-ocean-sea ice-land surface general circulation models (GCMs) that simulate climatic variability and change over decadal to centennial time scales and that simulate regional climate variability and change with greater fidelity. This effort will be closely coordinated with the DOE-wide Scientific Discovery through Advanced Computing (SciDAC) for Climate Change Research activities and will enhance partnerships with the Office of Science's Advanced Scientific Computing Research program. The focus will be on incorporation and testing of various aerosol schemes, convection schemes, ice sheets, and land-surface schemes in coupled models, and evaluation using innovative metrics that span a variety of climate time scales, specifically:

- Testing of newly developed convection schemes, cloud parameterization schemes, and global cloud-resolving models against observations, with emphasis on testing cloud-aerosol-radiation parameterization schemes in GCMs
- Characterizing aerosol-climate interactions, including testing and improving aerosol parameterization schemes in atmospheric GCMs
- Exploring decadal predictability of the climate system, and understanding cryospheric processes and their role in the climate system
- Understanding climate extremes in a changing climate
- Developing metrics for evaluation of climate models, including model diagnostics and intercomparison
- Developing new metrics for ocean model evaluation and diagnostics
- Developing and employing enabling technologies for climate model simulation dissemination.

Climate change information is being increasingly sought for impact studies and national and international assessments. These activities are at the interface of process research and global climate modeling, and are expected to accelerate process representation in coupled Earth system models for climate change projections. The DOE leadership class computational facilities now provide computing resources for models to be run at resolutions at which complex issues of data archival, management, and dissemination need to be addressed. DOE will develop such tools and capability.

DOE will continue projects initiated in FY 2008 on the topic of abrupt climate change. Also DOE's SciDAC for Climate Change Research will continue partnerships with the Advanced Scientific Computing Research program, including work towards the creation of a first-generation Earth System Model based on the Community Climate System Model which treats the coupling between the physical, chemical, and biogeochemical processes in the climate system. The model will include comprehensive treatments of the processes governing well-mixed greenhouse gases, natural and anthropogenic aerosols, the aerosol indirect effect, and tropospheric ozone for climate change studies. Research will develop and test a global cloud-resolving model using a geodesic grid, with grid-cell spacing of approximately 3 km, capable of simulating the circulation associated with large convective clouds.

Climate Forcing

Collection and analysis of data from DOE's Atmospheric Radiation Measurement (ARM) Cloud and Radiation Test Bed (CART) sites will continue in FY 2009 to improve understanding of the radiative transfer processes in the atmosphere and to formulate better parameterizations of these processes, especially cloud and aerosol effects for use in climate models. In FY 2009, ARM will complete the deployment of the ARM Mobile Facility (AMF) to China and will begin developing a second AMF for deployment in FY 2010. The AMF deployment to China will study the aerosol indirect effect. Aerosols in China have exceptionally high loading and diverse properties whose influence has been detected across the Pacific Rim. In FY 2009, the ARM science program will also focus on the development of new cloud schemes and improvement of cloud-radiation parameterization schemes. Special measurements from the Tropical Warm Pool International Cloud Experiment (TWP-ICE), the Convective and Orographically Induced Precipitation Study (COPS), the aerosol study in China, and the Cloud and Land Surface Interaction Campaign (CLASIC) will give scientists ample opportunities to validate and improve the representation of radiation-cloud-aerosol processes in climate models.

DOE's Atmospheric Science Program (ASP) will continue research in FY 2009 to reduce uncertainties in aerosol radiative forcing of climate. This research includes laboratory and field research on key processes individually and as encountered in "real world" environments. Acquired data are used to develop and test predictive parameterization schemes or models for aerosol properties and their effect on radiative transfer in the atmosphere. Field and laboratory observations are also used to interpret and extend the results of process model simulations. Current priority atmospheric processes under study include transformations and properties of carbonaceous aerosols, especially secondary organic aerosols that are poorly predicted by current atmospheric models. Also important are processes controlling new particle formation and growth, as well as the properties that affect their activation as droplet and crystal nuclei. During FY 2009, ASP will participate in a major collaborative interagency field campaign (VOCALS) aimed at measuring interactions of aerosols with clouds in a region that is affected both by pristine and polluted air masses. One specific objective of ASP activity is to test new process models of drizzle formation that show promise for inclusion into global climate models. Analysis of data from prior field studies will continue, principally from the FY 2006 campaign conducted in and around Mexico City to examine the properties and processes of aerosols emanating from a large metropolitan area, and from the FY 2007 Cumulus Humilis Aerosol Processing Study (CHAPS) campaign, that examined interactions of aerosols with fair-weather cumulus clouds.

DOE's Terrestrial Carbon Processes (TCP) research will continue to improve understanding of the role of terrestrial ecosystems in the global carbon cycle, with attention on processes that control the rate of carbon dioxide (CO_2) exchange with ecosystems and that affect the rate of atmospheric increase and climate forcing by this greenhouse gas. Research in FY 2009 will address the questions and elements described in Chapter 7 of the *Strategic Plan for the Climate Change Science Program*. TCP research will continue to contribute to the North American Carbon Program (NACP) through support of experiments, observations, and modeling of atmospheric CO_2 and the terrestrial carbon cycle. Research will continue to focus on the AmeriFlux network of observations, experiments, modeling, and syntheses. Temporal and spatial observations of gross and net CO_2 fluxes in several mid-latitude ecosystems in North America, and real-time information on ecosystem carbon states and sinks in these systems, will continue to be made available to researchers who are investigating regional CO_2 exchange, continental-scale carbon sinks and sources, and carbon cycle-climate relationships. DOE will also support the NACP strategy of a model-based comparison of "bottom-up" (distributed ecosystem models driven by land surface and meteorological information) and "top-down" (inferring spatially distributed surface fluxes from atmospheric measurements) approaches to estimating ecosystem CO_2 fluxes for different regions of the United States. As part of joint carbon cycle-climate change research to improve simulation models, DOE will provide information on biogeochemical and physiological responses and terrestrial ecosystem feedbacks related to climate change in several mid-latitude ecosystems of North America. Support will also be continued for the Carbon Dioxide Information and Analysis Center (CDIAC) to enable it to respond to data and information requests from users all over the world who have a need for data on, for example, greenhouse gas emissions and concentrations.

Climate Change Response

In FY 2009, DOE continues to sponsor experimental studies of the potential effects of warming, and changes in precipitation, on multiple terrestrial ecosystems. The new scientific data and understanding obtained by this research will facilitate informed decisionmaking about the means of producing the energy needed by society. It will do this by defining relationships between climatic changes that might be caused by energy production and the potential effects of those changes on the health of terrestrial ecosystems, and the organisms that they contain.

The primary focus in FY 2009 will be experimental studies of the potential effects of warming on the abundance and geographic distribution of plant and animal species in several ecosystem types. The experiments will be conducted to fill specific critical knowledge gaps. In particular, experiments will determine linkages between warming and the possibility of species migrations, the expansion of species into areas that are presently too cool for their success, and the decline of species or ecosystems presently at the warm edge of their ranges. Field experiments will be conducted in high-elevation forests and meadows associated with the alpine tree line, the transition zone (ecotone) between temperate and boreal forests, and western shrubland. In addition to field experiments, laboratory experiments will determine relationships between warming and the success of plants and animals in model ecosystems. Laboratory studies will focus on key testable hypotheses about ecological effects of warming.

DOE will continue to provide core support for operation of the world's largest long-term field study of the potential effects of changes in atmospheric composition caused by energy production on a terrestrial ecosystem. That experiment (in central Wisconsin) is enriching the atmosphere within forest communities with CO_2 and ozone (O_3) concentrations that are anticipated to occur within 50 years. The experiment is documenting direct and indirect effects of elevated CO_2 and/or O_3 on three tree species, soil microorganisms, and pests that feed on trees.

The Integrated Assessment of Global Climate Change Research Program will continue to support research on the nature and magnitude of human-Earth systems interactions, providing scientific insights into the integrated drivers of climate change and the impacts of and adaptations to those changes. The program will deliver improved science-based tools for determining safe levels of greenhouse gas emissions and understanding of the relative efficiencies and impacts of potential mitigation strategies. Consistent with recommendations by the Biological and Environmental Research Advisory Committee, the research will undergo a transformation and will shift considerable attention to the challenge of representing climate change impacts and adaptations within integrated assessment models. Development of non-monetary valuation and visualization methods and tools will be an important dimension of this new work. Additionally, DOE will explore the application of more advanced computational platforms reflecting the need for tighter coupling between what are presently reduced-form models and the rich detail and reduced uncertainty of underlying biogeophysical models.

DOE will also continue support of its Global Change Education Program in FY 2009, including support of undergraduate and graduate students through the DOE Summer Undergraduate Research Experience (SURE) and the DOE Graduate Research Environmental Fellowships (GREFs).

Related Research

DOE plays a major role in carbon sequestration research to reduce atmospheric concentrations of energy-related greenhouse gases, especially CO_2, and their net emissions to the atmosphere. The research builds on, but is not part of, the CCSP. It focuses on both developing the scientific information needed to enhance the natural sequestration of excess atmospheric CO_2 in terrestrial systems, and assessing the potential environmental consequences and ancillary benefits of that enhanced sequestration. It also includes research to develop biological approaches for sequestering carbon either before or after it is emitted to the atmosphere. Funding for DOE's carbon sequestration research is part of the Climate Change Technology Program (CCTP). CCTP also provides related research funding to support a balanced and diversified portfolio of advanced technology research and development, focusing on energy-efficiency enhancements; low greenhouse gas emission energy supply technologies; carbon capture, storage, and sequestration; and technologies to reduce emissions of non-CO_2 gases. Together, CCSP and CCTP will help lay the foundation for future progress. Advances in the climate change sciences under CCSP can be expected to improve understanding about climate change and its impacts. Similarly, advances in climate change technology mitigation under CCTP can be expected to bring forth an expanded array of advanced technology options at a lower cost which will help reduce greenhouse gas emissions.

DEPARTMENT OF HEALTH AND HUMAN SERVICES

National Cancer Institute (NCI)
National Institute of Arthritis and Musculoskeletal and Skin Diseases (NIAMS)
National Institute of Environmental Health Sciences (NIEHS)

Principal Areas of Focus

The Department of Health and Human Services (HHS) supports a broad portfolio of research related to environmental health and the health effects of global change. The National Institutes of Health (NIH) supports CCSP research that focuses on exposure to ultraviolet (UV) and near-UV radiation, with the principal objectives being an increased understanding of the effects of UV and near-UV radiation exposure on target organs (e.g., eyes, skin, immune system), the molecular changes and genetic susceptibilities that lead to these effects, and the development of strategies to prevent initiation or promotion of disease before it is clinically defined.

In addition to UV and near-UV radiation research, HHS also supports other research related to the health effects of global change. For example, NIEHS supports research on the health effects of air pollution and temperature, agricultural chemicals, and materials used in new technologies to mitigate or adapt to climate change. In addition, the Centers for Disease Control and Prevention (CDC) is engaged in a number of activities related to climate change, such as emerging and reemerging infectious diseases. Such related research is growing in importance.

Program Highlights for FY 2009

The NIEHS program supports grants and intramural projects that investigate the effects of UV exposure on the immune system, aging process, sensitive tissues such as the retina and skin, and methods to reduce these harmful effects. Examples of research include projects based on the premise that solar radiation contributes to development of melanoma by inducing chromosomal damage in skin cells. Malignant melanomas are of significant public health concern because their incidence is rising and no effective medical intervention is available for reducing morbidity and mortality. New findings from these studies may lead to the discovery of biomarkers with potential therapeutic and predictive value for specific types of melanomas and different stages of melanoma progression. The National Toxicology Program (NTP) funded and operated by NIEHS is carrying out a systematic analysis of commercially available sunscreens to characterize several nanoscale metal oxides (e.g., titanium, zinc) currently used with regard to their dermal penetration and photocatalytic action. Careful attention is being paid to determining critical aspects of size, surface area and chemistry, crystallinity, and biopersistence in relation to both dermal penetration and potential for toxicity in the presence or absence of simulated solar light.

NCI is supporting a wide range of studies to characterize the etiology, biology, immunology, and pathology of a variety of changes in the skin (morphological effects that might precede skin cancer), including photoaging, non-melanoma skin cancers, and melanoma caused by exposure to UV radiation. In addition, NCI supporting studies to reduce the risk of melanoma and non-melanoma skin cancer through the development of clinically useful primary and secondary prevention strategies. One study is developing, implementing, and evaluating solar protection programs for middle school children. The interventions target school, community, recreation and beach settings, primary care practices, and parents. The interventions are based on theories that include social influence, psychological factors, and cognitive decisional factors in adolescence. Other studies are looking at the role of UV light exposure in the development of second malignant neoplasms in cancer survivors.

NIAMS supports basic and clinical research on the effects of UV-A and UV-B radiation on skin. Examples of current studies include research on the mechanisms by which UV light induces pigmentation in the skin and the potential role of pheomelanin in the development of melanomas in fair-skinned individuals. These studies may lead to the development of "sunless tanning" products that achieve the protective effects of a tan without exposure to UV light. Other studies look at the role of different genes in the acute and long-term response of epidermal keratinocytes to UV exposure, including skin carcinogenesis and the maintenance of stem cell populations. Another study examines the role of the PPARgamma pathway in UV-B stress response and photocarcinogenesis. Several PPARgamma agonists are currently in use for the treatment of type II diabetes and may have chemopreventive activity. Another study looks at the effects of UV-B radiation on the stability of cell cycle regulatory proteins, yielding insight into the mechanisms by which UV-B radiation increases the risk of non-melanoma skin cancer. There are also studies that are testing the role of acquired homoplasmic mitochondrial changes in the process of photoaging and aberrant keratinocyte hyperplasia, novel biosynthetic pathway for secosteroids (such as vitamin D) in the skin, as well as the expression of keratins induced whenever skin tissue is subjected to injury, UV exposure, and other challenges. A patient-oriented research project involves the molecular mechanisms for the exaggerated response to UV-B of a polymorphism that is strongly associated with a photosensitive form of lupus erythematosus. Research is also conducted on the effect of UV-R on Langerhans cells, star-shaped cells in the germinative layer of the epidermis, and on immunity in skin. Using gene array technology, scientists have identified 52 genes that are consistently up-regulated by UV-R.

Related Research

CDC conducts public health research on a wide variety of topics that are associated with climate change, ranging from vector-borne diseases to human health effects of heat waves. CDC has established a long-term national surveillance system to monitor enzootic transmission activity and patterns across the entire country. For West Nile virus (WNV), the agency is conducting research on the potential human health burdens and transmission characteristics of the disease in Guatemala to study the ecology of WNV and other arboviruses causing encephalitis. The results of the ecology studies may lead to a better understanding of how climate change may influence transmission dynamics. CDC is conducting interrelated investigations of the complex ecology of WNV to better understand its distribution in the United States. Scientists from CDC are working with colleagues from around the world to analyze the key climatic variables and other ecological factors that impact the transmission and distribution of other zoonotic diseases including Chikungunya viral fever, Japanese encephalitis, Rift Valley Fever, and plague. Researchers are developing mathematical models that relate changing weather conditions, among other factors, to the risk of infectious diseases in humans, including those caused by hantavirus, lyssaviruses, and filoviruses.

CDC is also developing models to predict mortality risks from the most direct effects of climate change—heat waves. Collaborations with university colleagues on four current projects use remote-sensing data to determine urban neighborhoods and populations most at-risk for deaths during an extreme heat event. Research is also focusing on the knowledge, attitudes, and beliefs of the public on health issues related to climate change to effectively craft health education messages.

DEPARTMENT OF THE INTERIOR

Principal Areas of Focus

Department of Interior (DOI) / U.S. Geological Survey (USGS) research and observations contribute directly to CCSP strategic goals, principally through integrated multidisciplinary data-collection networks and studies designed to understand the interactions between climate, earth surface processes, and ecosystems on time scales ranging from years to millennia. By combining the expertise of hydrologists, geologists, biologists, geographers, and remote-sensing scientists within one organization, USGS supports truly interdisciplinary research and assessment of trends in resource condition in the following major focus areas:

- Studies of climate history and impacts on landscapes and ecosystems
- Hydrologic impacts of climate change
- Carbon cycle science
- Land-use and land-cover changes
- Decision-support research and development.

The goal of global change research at USGS is to improve knowledge and understanding of the Earth's past and present climate and environment, the forces bringing about changes in the Earth's climate, and the sensitivity and adaptability of natural and managed ecosystems to climate changes.

Program Highlights for FY 2009

USGS is beginning a focused effort to develop decision-support tools for policymakers and resource managers to assess our ability to cope with and adapt to the various effects of climate change. The USGS Climate Effects Science Network (CESN)—coordinated through all DOI resource management Bureaus—will integrate climate- and environmental-change data sets with conceptual and digital models across disciplines including remote sensing, geography, geology, biology, and hydrology to better understand impacts to natural resources, agriculture, and human populations on episodic to decadal and millennial time scales, local to global spatial scales, and weather to climate process scales. The goal of the network is to develop a systems-level understanding of biogeochemical processes resulting from changes in climate, to link these changes to the sustainability of ecosystems, wildlife, subsistence cultures, and societal infrastructure, and to apply the knowledge gained for decision support. USGS is in a unique position in the Earth science research and applications community because of its ability to leverage and integrate research and monitoring results across the Earth system science disciplines with *in situ* data collection capability, space-based and airborne observational platforms, high-end computing capabilities, data and information management systems, and decision-support tool development.

Monitoring, understanding, assessing, and predicting changes in Earth processes and the associated impacts to decisionmaking will be a central focus of CESN during its initial development within the Global Change Program and will continue as the focus for the network through 2009 and beyond. For example, individual USGS programs and projects such as the MIT-USGS Science Impact Collaborative (MUSIC) and the Policy Analysis and Science Assistance Branch of USGS's Fort Collins Science Center are developing decision-support tools from monitoring data sets, research results, and model projections that are products of the network. In 2008, USGS initiated a pilot integrated research area in northern Alaska, where permafrost thaw and sea-ice melting are resulting in rapid and poorly understood changes to regional ecosystems. In 2009, USGS will expand collaborations with the above-mentioned and similar programs across the Federal, State, and academic sectors to develop a nationally integrated network that will provide state-of-the-art environmental observations, research, and applications. The main objective of this initiative will be the development of cost-effective strategies for adaptation to and mitigation of the effects of climate change through the creation of science-based decision-support tools and, in the longer term (next 2-3 years, 2009-2011), decision-support systems.

USGS Global Change Programs

Earth Surface Dynamics

The Earth Surface Dynamics (ESD) program is USGS's primary provider of scientific information on past climates and their implications for Earth and human systems. ESD has the following specific research objectives:

- Document the nature of climatic and environmental change and variability on time scales ranging from years to millennia
- Develop fundamental understanding of interactions between climate, earth surface processes, and marine and terrestrial ecosystems on time scales ranging from years to millennia
- Seek to understand impacts of climate change and variability on landscapes and marine and terrestrial systems
- Model and anticipate the effects of climate change and variability on natural and human systems
- Provide information on the relative sensitivity, adaptability, and vulnerability of ecosystems, resources, and regions to climatic change and variability to support land and resource management and policy decisions.

Research to understand Earth surface processes and climate change impacts provides both rigorous background science and perspectives on consequences for policymakers and for use by land and resource managers. Research activities include projects that study past climates and environments, investigate the effects of past and present climate change and climate variability on landscapes and ecosystems, study the processes involved in landscape change, and forecast the effects of climate change and variability on landscapes and ecosystems.

Investigations of climate and environmental history are conducted in a wide range of locations to ensure that the program provides policy-relevant information on past climatic conditions for a wide range of ecosystems and landscapes. Projects range in location from the North Slope of Alaska to the southwestern United States, from the Mississippi Valley to the Everglades, and from the Channel Islands to the Chesapeake Bay. Many of these projects provide information on such factors as land use, erosion and sedimentation, and thawing permafrost, and the implications and impacts of these to directly support land and resource management decisions. Partners and customers include Federal land management agencies (including the Fish and Wildlife Service, National Parks Service, the U.S. Forest Service, and the Bureau of Land Management), other Federal agencies (including NOAA and NSF), State governments and consortia, international partners (including the Canadian government and Canadian federal agencies), universities and university scientists, tribal organizations, and non-governmental organizations such as the Chesapeake Bay Program.

Geographic Analysis and Monitoring

Research is directed to the understanding of the rates, causes, and consequences of landscape change over time. This knowledge is used to model processes of landscape change and to forecast future conditions. Studies are designed to document and understand the nature and causes of changes occurring on the land surface; to assess the impacts of land surface changes (including urbanization) on ecosystems, climate variability, biogeochemical cycles, hydrology, and human health; and to develop the best methods to incorporate science findings in the decisionmaking process.

Hydroclimatology

Research on effects of climate change and variability on the hydrologic cycle focuses on characterizing, and developing predictive methods related to, the hydroclimatology of North America. This includes identification of seasonal variations in regional streamflow in relation to atmospheric circulation (for regional streamflow prediction and flood/drought hazard assessment); the linkage between atmospheric circulation and snowpack accumulation (for forecasting spring and summer water supply in the western United States and for flood forecasting), as well as glacier mass balance; and the physical and chemical variability in riverine and estuarine environments in relation processing occurring in their contributing watersheds, and to large-scale atmospheric and oceanic conditions (to discriminate natural from human-induced effects on such systems). It also includes documenting the long-term behavior of hydrologic systems in response to past climatic variations and changes (from decades to hundreds of thousands of years) as well as more recent (decadal) hydrologic trends. The program maintains an active effort to develop improved representations of terrestrial hydrologic processes in general circulation and regional climate models. In broad terms, these activities are aimed at improving statistical and deterministic methods for predicting hydrologic hazards and related environmental conditions on monthly to interannual time scales.

Carbon Cycle

USGS conducts a broad range of carbon cycle research focused on North America, which includes:

- *Assessment of Carbon Stocks and Soil Attributes*—Determining the spatial distribution of carbon in the terrestrial environment in relation to historical natural and human processes, as a basis for initializing dynamic models of soil carbon. Measuring soil chemistry has focused on the Mississippi and Delaware River basins, the latter in collaboration with the USDA Forest Service Forest Inventory and Analysis Program.
- *Carbon Sequestration in Sediments*—Studying the re-deposition of eroded soils and sediments (and their associated organic carbon) which sequesters large quantities of carbon, buried at the base of slopes and in wetlands, riparian areas, and reservoirs.
- *Carbon Sequestration in Wetlands*—Field and laboratory process studies, spatial analysis, and modeling are being used in wetlands of the Lower Mississippi River Valley and the Prairie Pothole Region to quantify the influence of land-use change on carbon sequestration and greenhouse gas emissions and to identify environmental factors controlling carbon sequestration. These studies will provide recommendations and decision-support tools to resource managers to maximize carbon sequestration benefits consistent with DOI goals for restoration of ecosystem services such as habitat, flood storage, and water quality.
- *Landscape Dynamics and Vegetation Change*—Examining the long-term dynamics of vegetation change in relation to climate change and variability. A detailed history of vegetation change in the western United States is being constructed. Past changes are used to model vegetation response to climatic variables. This knowledge is applied in forecasting the effects of future climate change on the distribution of vegetation in the western United States.
- *Fate of Carbon in Alaskan Landscapes*—Expanding process studies and modeling to better understand both the historic and modern interactions among climate, surface temperature and moisture, fire,

and terrestrial carbon sequestration. Cold region forests (boreal ecosystems) contain large carbon reserves that are highly susceptible to changes in climate.

- *Exchanges of Greenhouse Gases, Water Vapor, and Heat at the Earth's Surface*—Employs field measurements, remote sensing, and modeling of carbon fluxes to develop estimates of gross primary productivity, respiration, and net ecosystem exchange at flux tower sites, and uses remotely sensed data to extrapolate these carbon fluxes to ecoregions.

Changes in Ecosystems

USGS global change research on ecosystems aims to determine the sensitivity and response of ecosystems and ecological processes to environmental factors, including existing climate and natural and anthropogenic impacts, at the local, landscape, regional, and continental level; to assess and predict how future environmental conditions may affect the structure, function, and long-term viability of natural and human-impacted ecosystems; and to provide scientific knowledge and technologies needed for conservation, rehabilitation, and management of sustainable ecosystems. Current USGS ecosystems research focuses on:

- The relative sensitivity of biological resources and geographic areas of the Nation to global changes in order to detect early changes and to prioritize action
- The causal mechanisms underlying ecosystem responses to global change
- The role of scaling in understanding and managing the spatial and temporal responses of biological systems to global change
- Development and testing of management options for adapting to the effects of global change and minimizing undesired effects of global change.

Satellite Data Management and Dissemination

USGS operates and continually enhances the capabilities of the Center for Earth Resources Observation System (EROS) to serve as the National Satellite Land Remote-Sensing Data Archive, by maintaining existing data sets, adding new ones, and converting older data sets from deteriorating media to modern, stable media. The archive's holdings are used for environmental research, land management, natural hazard analysis, and natural resource management and development with applications that extend well beyond U.S. borders. The worldwide community of archive users includes personnel in Federal, State, local, and tribal governments, researchers at academic institutions, private enterprise, and the public.

Land Use and Land Cover

The Land Cover Characterization Program was started in 1995, to address national and international requirements for land-cover data that were becoming increasingly sophisticated and diverse. The goal is to be a national and international center for excellence in land-cover characterization, via:

- Development of state-of-the-art multiscale land-cover characteristics databases used by scientists, resource managers, planners, and educators (global and national land cover)
- Contribution to the understanding of the patterns, characteristics, and dynamics of land cover across the Nation and the Earth (urban dynamics and land-cover trends)
- Pursuit of research that improves the utility and efficiency of large-area land-cover characterization and land-cover characteristics databases
- Serving as a central facility (Land Cover Applications Center) for access to, or information about, land-cover data.

DEPARTMENT OF STATE

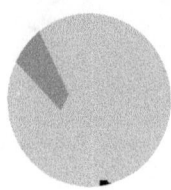

Principal Areas of Focus

Through Department of State (DOS) annual funding, the United States is the world's leading financial contributor to the United Nations Framework Convention on Climate Change (UNFCCC) and to the Intergovernmental Panel on Climate Change (IPCC)—the principal international organization for the assessment of scientific, technical, and socioeconomic information relevant to the understanding of climate change, its potential impacts, and options for adaptation and mitigation. Recent DOS contributions to these organizations provide substantial support for global climate observation and assessment activities in developing countries. DOS also works with other agencies in promoting international cooperation in a range of bilateral and multilateral climate change initiatives, including: the Asia-Pacific Partnership on Clean Development and Climate, the Carbon Sequestration Leadership Forum, the Group on Earth Observations, the Methane-to-Markets Partnership, the International Partnership for a Hydrogen Economy, and the Generation IV International Forum.

Program Highlights for FY 2009

During FY 2009, DOS will continue to support the activities of the UNFCCC and the IPCC, and will advance the bilateral and multilateral partnerships for global climate science, technology, and observation that were undertaken in the FY 2006-2008 time frame.

DEPARTMENT OF TRANSPORTATION

Principal Areas of Focus

The Department of Transportation (DOT) conducts research and uses existing science to improve decisionmaking tools to address climate change. DOT supports research that 1) examines the potential impacts of climate variability and change on transportation infrastructure and services, 2) increases energy efficiency and reduces greenhouse gases, and 3) improves transportation greenhouse gas data and modeling. DOT has many programs that have either direct or indirect climate benefits, and is working to develop cross-modal strategies to reduce greenhouse gas emissions.

DOT's Climate Change Center is the Department's focal point for information and technical expertise on climate change. The Center coordinates research, policies, and actions related to transportation and climate change with DOT's component organizations. Supporting DOT's core goals of safety, mobility, environmental stewardship, and security, the Center promotes comprehensive approaches to reduce greenhouse gases, to prepare for the potential impacts of climate change, and to develop necessary adaptations to transportation operations and infrastructure. The Center's three primary objectives follow:
*   Promote cost-effective strategies that reduce greenhouse gas emissions while supporting improved transportation safety, mobility, and efficiency
*   Foster strategies to avoid, mitigate, or adapt to the potential impacts of climate change on the transportation system
*   Provide leadership to the transportation community and coordinate DOT multi-modal activities on climate change.

The Center supports CCSP goals through these objectives. Specifically, the Center aims to inform CCSP Goal 4 by identifying and providing scientific inputs for evaluating adaptation options and CCSP Goal 5 by supporting adaptive management and planning for physical infrastructure sensitive to climate variability and change.

In addition to participating in the Center, the Federal Aviation Administration (FAA) has programs to assess and identify potential measures to reduce fuel consumption and greenhouse gas emissions. FAA conducts research to support CCSP Goal 2, leveraging research with other U.S. Government agencies to reduce uncertainties surrounding aviation emissions and their effect on climate change. For example, FAA research through the Partnership for Air Transportation Noise and Emissions Reduction (PARTNER) Center of Excellence addresses the impact of aircraft contrails on climate change. FAA also has a number of ongoing operational initiatives to reduce fuel consumption and thus the amount of greenhouse gas emissions produced by aviation, including improved air traffic management, reduced vertical separation minimums, and the voluntary airport low emissions program that assists in deploying low emissions technology to airport operations. FAA also participates heavily in the work program of the International Civil Aviation Organization's Committee on Aviation Environmental Protection, and provides technical expertise and data to the Intergovernmental Panel on Climate Change (IPCC) and the United Nations Framework Convention on Climate Change (UNFCCC).

Program Highlights for FY 2009

DOT's Climate Change Center is undertaking several research projects that support CCSP Goals 4 and 5:
- Refining a tool to allow comparative analysis of emissions from different modes of transportation, including aviation, automobile, marine, and diesel transportation
- Exploring adaptation to potential impacts of climate change by partnering with the Transportation Research Board of the National Academies
  - Reexamine the role of design standards for transportation infrastructure in light of potential impacts from climate change
  - Develop operational responses to potential climate change impacts
  - Review approaches to decisionmaking under uncertainty
- Conducting an emissions analysis of freight transport, comparing land-side and water-side short-sea routes to develop and demonstrate a decision modeling tool
- Determining the potential effects of sea-level rise on national transportation infrastructure
- Preparing a study for Congress of the impact of passenger and freight transportation on climate change and solutions
- Developing a web clearinghouse of passenger and freight solutions across all modes to reduce greenhouse gas emissions and environmental impact
- Conducting a pilot project to record best practices for metropolitan planning organizations (MPOs) to incorporate climate change into transportation planning.

The Center sponsored CCSP Synthesis and Assessment Product 4.7, *Impacts of Climate Variability and Change on Transportation Systems and Infrastructure–Gulf Coast Study*. This project—initiated under the President's Climate Change Research Initiative—was a joint research effort with USGS. A Federal Advisory Committee was formed in 2006, and Phase I was completed in 2007. Phase I provided an integrated overview of infrastructure sensitivities in the region. The final document was released in March 2008.

The Office of the Secretary is funding several projects, including a *Best Practices Guidebook for Greenhouse Gas Reductions in Freight Transportation* designed for use by companies and individual freight operators.

Related Research

Many of DOT's programs have ancillary climate benefits:
- The Federal Highway Administration has numerous programs to prepare the highway system for weather irregularities and to reduce air pollutants:
  - *Road Weather Management Program.* This program seeks to better understand the impacts of weather on roadways. The Clarus initiative will develop and demonstrate a national integrated surface transportation road weather observing, forecasting, and data management system.
  - *Congestion Mitigation and Air Quality (CMAQ) Improvement Program.* The CMAQ program provides over $8.6 billion in funds over a period of 5 years (2005-2009) to State transportation agencies to invest in projects that reduce emissions from transportation-related sources.
  - *Idle-Reduction Activities.* DOT, EPA, and DOE have provided funding for the implementation of idle-reduction projects around the country (both on-board and off-board technologies) for transportation, air quality, and energy stakeholders. The projects have resulted in reductions in

criteria air pollutants, such as nitrous oxides, as well as reductions in carbon dioxide emissions. This initiative has expanded to include idling emissions from marine, rail, and off-road heavy-duty engines.

- The National Highway Traffic Safety Administration sets new Corporate Average Fuel Economy standards for light trucks, increasing energy efficiency and thus decreasing greenhouse gas emissions. In FY 2008, NHTSA is developing café regulations for future model years, including implementing provisions of the Energy Independence and Security Act.

- FAA continues to develop a comprehensive suite of software tools to allow for the thorough assessment of the environmental effects and impacts of aviation (<http://www.faa.gov/about/office_org/headquarters_offices/aep/models>). One element of the tools suite is the Aviation Environmental Design Tool (AEDT), which incorporates the legacy System for assessing Aviation's Global Emissions (SAGE) tool. AEDT/SAGE takes as input detailed fleet descriptions and flight schedules, and produces estimates of noise and emissions inventories and fuel consumption at global, regional, and local levels. The tool can also estimate future output based upon forecasts, including potential technology advances or operational improvements, as well as the influence of potential market-based measures to reduce fuel consumption and greenhouse gas emissions. Data from AEDT/SAGE is used to calculate the FAA's Flight Plan aviation fuel efficiency goal.

- The Federal Transit Administration (FTA) Fuel Cell Program has researched and demonstrated fuel cell bus technology since the mid-1990s. FTA also conducts alternative fuels research.

- DOT's Research and Innovative Technology Administration conducts and administers a range of alternative fuel grants. The research focuses on developing and demonstrating alternative fuel and advanced propulsion technology systems with an emphasis on efficiency, safety, and sustainability. Research aims to inform stakeholders and facilitate technology.

- Other programs for congestion mitigation, hydrogen-powered transportation, and transit developments all will potentially reduce greenhouse gases.

AGENCY FOR INTERNATIONAL DEVELOPMENT

Principal Areas of Focus

The Famine Early Warning System Network (FEWS NET) is an innovative application of science for supporting efforts to alleviate risks related to existing climate variability or potential climate change. Through FEWS NET, the U.S. Agency for International Development (USAID) is able to provide decisionmakers—both in the United States and in the developing world—with information designed to support policy and program interventions for effective and timely response to drought and food insecurity. FEWS NET historically focused its activities in 23 countries of Africa, as well as Guatemala, Haiti, and Afghanistan. In addition to monitoring a wide variety of socioeconomic indicators to identify levels of food insecurity, FEWS NET monitors and analyzes remotely sensed data and ground-based meteorological, crop, and rangeland observations to track the progress of rainy seasons and crop production in semi-arid regions, in order to identify early indications of reduced food availability and access.

Program Highlights for FY 2009

In FY 2009, FEWS NET will continue to provide seasonal monitoring in relationship to analyses of food insecurity conditions in Africa and Central America, as well as in Haiti and Afghanistan. A major element of this work has been to strengthen information networks that collect and analyze data to reveal intra- and interannual climate variability trends as they relate to possible longer term climate variability and change.

Depending upon available funding, USAID will continue to conduct climate change adaptation programs applying scientific information to international development assistance.

ENVIRONMENTAL PROTECTION AGENCY

Principal Areas of Focus

The core purpose of the Global Change Research Program in the U.S. Environmental Protection Agency (EPA) Office of Research and Development is to provide scientific information to stakeholders and policymakers to support them as they decide whether and how to respond to the risks and opportunities presented by global change. The program is stakeholder-oriented, with primary emphasis on assessing the potential consequences of global change (particularly climate variability and change) on air quality, water quality, aquatic ecosystems, and human health in the United States. The program's focus on these four areas is driven by EPA's mission and statutory and programmatic requirements. EPA uses the results of these studies to investigate adaptation options to improve society's ability to effectively respond to the risks and opportunities presented by global change, and to develop decision-support tools for resource managers coping with a changing climate. EPA also has begun to invest in decision-support tools to help decisionmakers evaluate alternative strategies for reducing greenhouse gas emissions and the environmental implications of those strategies.

The program uses a place-based approach because the impacts of global change and their solutions are often unique to a location (e.g., a watershed). Partnerships are established with locally based decisionmakers to ensure that the program is responsive to their unique scientific information needs and the socioeconomic realities at their locales.

EPA's work is consistent with and closely coordinated with the *Strategic Plan for the Climate Change Science Program*. Planning and implementation of EPA's program is integrated with other participating Federal departments and agencies to reduce overlaps, identify and fill programmatic gaps, and add value to products and deliverables produced under the CCSP's auspices. EPA coordinates with other CCSP agencies to develop and provide timely, useful, and scientifically sound information to decisionmakers. EPA is committed to support of CCSP's research and assessment activities. This commitment includes assessments uniquely focused on EPA's mission and statutory requirements (e.g., assessments of the impacts of global change on air and water quality) and support for statutory mandates on the CCSP to produce periodic assessments of the potential impacts of climate change for Congress. Also, as called for by the National Research Council in 2001, EPA supports and fosters projects that link the producers and users of knowledge in a dialog that builds a mutual understanding of what is needed, what can credibly be said, and how it can be said in a way that maintains scientific credibility.

EPA's program has two major areas of emphasis: air quality and water quality/aquatic ecosystems. The program also evaluates the human health consequences of the changes in air quality, water quality, and aquatic ecosystems.

Air Quality

Studies are underway that examine the potential consequences of global change on air quality in the United States. The long-term goal of this focus area is to provide the approaches, methods, and models to quantitatively evaluate the effects of global change on air quality, and to identify technology advancements and adaptive responses and quantify their effect on air quality.

Water Quality / Aquatic Ecosystems

EPA's mission is to protect human health and safeguard the natural environment. EPA provides environmental protection that contributes to making communities and ecosystems diverse, sustainable, and economically productive. Consistent with this goal, EPA's Global Change Research Program is assessing the impacts of global change on water quality and aquatic ecosystems in the United States.

Water quality is affected by changes in runoff following changes in precipitation and evapotranspiration and/or changes in land use. The program is investigating the possible impacts of global change (climate and land-use change) on water quality using a watershed approach. A major focus is on studying the sensitivity to climate change of goals articulated in the Clean Water Act and the Safe Drinking Water Act, and the opportunities available within the provisions of these Acts to address anticipated impacts.

The program also has planned research activities that evaluate the effects of global change on aquatic ecosystems (which may include lakes, rivers, and streams; wetlands; and estuaries and coastal ecosystems), invasive non-indigenous species, and ecosystem services. EPA's investigations of the effects of global change on aquatic ecosystems uses as input the research being done by other CCSP agencies on marine and terrestrial ecosystems. Therefore, EPA's ability to successfully complete its assessments depends crucially upon the ability of other CCSP agencies to complete their related research activities.

Human Health

Since health is affected by a variety of social, economic, political, environmental, and technological factors, investigating the health impacts of global change is a complex challenge. As a result, health studies in EPA's Global Change Research Program go beyond basic epidemiological research to develop integrated health evaluation frameworks that consider the effects of multiple stresses, their interactions, and human adaptive responses. Along with health sector studies conducted in conjunction with other CCSP agencies, there are research activities focused on the possible consequences of global change on weather-related morbidity and vector- and water-borne diseases. In addition, the results from the program's air quality studies and water quality studies will be used to evaluate health consequences.

Intramural and extramural research contribute to all of EPA's investigations. In an attempt to capitalize on expertise in the academic community, a significant portion of the program's resources is dedicated to extramural research grants administered through the STAR (Science to Achieve Results) program. The STAR program focuses on science to support investigations of the consequences of global change for air quality, ecosystems, and human health in the United States. EPA will continue to coordinate closely with other CCSP agencies to identify the specific topics that should be emphasized within the STAR program.

Program Evaluation

The EPA Global Change Research Program is evaluated through extensive review by EPA's independent Board of Scientific Counselors (BOSC). A review in 2006 by the BOSC concluded that the program has conducted the "right work" and done it "well." The program "has provided substantial benefits to the Nation" and "is on course to make significant further contributions to societal outcomes by informing and facilitating decisions by the public and private sector actors who must consider the prospects of global change."

Program Highlights for FY 2009

EPA will continue to make significant contributions to the ongoing research activities of CCSP, and provide timely and useful information to resource managers coping with a changing climate. EPA-sponsored investigations will continue to be conducted through public-private partnerships that actively engage researchers from the academic community, decisionmakers, resource managers, and other affected stakeholders. Highlights of specific activities to be completed by or initiated in FY 2009 follow:

- Complete the three CCSP synthesis and assessment products for which EPA is the Lead Agency
- Initiate the second phase of EPA's quantitative assessment of the effects of global change on air quality in the United States
- Release final report on the potential impacts of climate change on combined sewer overflow mitigation in the Great Lakes and New England Regions
- Initiate an assessment of the impacts of climate change on water quality in the United States in support of EPA's statutory requirements under the Clean Water Act and the Safe Drinking Water Act
- Release an assessment of the effects of climate change and interacting stressors on the establishment and expansion of aquatic invasive species, and the implications for resource management
- Complete an assessment of the consequences of global change for water quality related to biocriteria
- Release a new online Climate Assessment Tool that provides resource managers with the ability to assess and manage impacts of climate change on sediment loadings to streams (e.g., through the use of riparian buffer zones)
- Co-sponsor with NOAA a study by the National Research Council of strategies and methods for climate-related decision support
- Issue a joint Request for Proposals with the Centers for Disease Control and Prevention focusing on the potential impacts of climate change on human health in the United States.

Related Research

In addition to focused CCSP activities, EPA conducts research that contributes to the characterization and understanding of risks to ecosystems and to human health. The ecosystems-based research is designed to understand and predict ecosystem exposure, responses, and vulnerabilities to high-risk chemicals and non-chemical stressors (e.g., invasive species, genetically altered organisms) at multiple scales of biological organization and geographic scales. The research in human health is oriented toward assessing the cumulative health risks to humans (e.g., cancer, reproductive, cardiovascular)—including high-risk subpopulations (e.g., children)—from chemical stressors emanating from multiple sources. Both of these major research areas will be affected by and are inextricably interrelated with climate change.

NATIONAL AERONAUTICS AND SPACE ADMINISTRATION

Principal Areas of Focus

The National Aeronautics and Space Administration (NASA) conducts a program of breakthrough research to advance fundamental knowledge on the most important scientific questions on the global and regional integrated Earth system. NASA's program encompasses all themes of the *Strategic Plan for the U.S. Climate Change Science Program*. NASA's investment in the 13-agency CCSP is 58% of the total amount of the President's 2009 Budget Request for CCSP. In addition, NASA contributes substantially to other interagency initiatives complementary with CCSP, including the *U.S. Ocean Action Plan* and the *Strategic Plan for the U.S. Integrated Earth Observation System*.

NASA continues to enhance its worldwide leadership in interdisciplinary science of the global integrated Earth system. The research encompasses the global atmosphere; the global oceans including sea ice; land surfaces including snow and ice; ecosystems; and interactions between the atmosphere, oceans, land and ecosystems, including humans. NASA's goal is to understand the changing climate, its interaction with life, and how human activities affect the environment. In association with national and international agencies, NASA applies this understanding for the well-being of society.

NASA presently operates 14 on-orbit satellites: ACRIMSAT, Aqua, Aura, CALIPSO, CloudSat, EO, GRACE, ICESat, Jason, Landsat-7, QuikSCAT, SORCE, Terra, and TRMM. Table 1 lists primary CCSP themes of the operating missions. NASA has seven missions in development (Aquarius, Glory, GPM, LDCM, NPP, OCO, and OSTM) for launch from 2008 to 2013, with OSTM, OCO, and Glory planned for launch in the 2008-2009 period. The President's 2009 Budget Request contains National Research Council (NRC) *Decadal Survey* recommendations, including the start of two new missions (ICESat-II and SMAP), pre-formulation activities on two additional missions, and initiating competitive Venture-class mission activities to ensure flexibility and community input into the evolving satellite constellation.

NASA aircraft- and surface-based instruments are used to calibrate and enhance interpretation of high-accuracy, climate-quality, stable satellite measurements. NASA supports state-of-the-art computing capability and capacity for extensive global integrated Earth system modeling. NASA, in recording approximately 4 terabytes of data every day, maintains the world's largest scientific data and information system for collecting, processing, archiving, and distributing Earth system data to worldwide users.

Program Highlights for 2009

NASA will make significant progress in 2009 in every theme of the *Strategic Plan of the U.S. Climate Change Science Program*. The examples below also demonstrate progress in all nine CCSP priorities for 2009.

Atmospheric Composition
NASA, in pioneering the understanding of the role of atmospheric chemistry and composition in the global integrated Earth system, adopted an overall research approach of sustained, systematic satellite observations with laboratory, aircraft, and ground-based measurements. Field campaigns are an important extension of satellite measurements.

CORRELATION OF NASA OPERATING SATELLITE MISSIONS WITH CCSP SCIENTIFIC THEMES

Satellite	Launch Date	CCSP Science Focus Areas
ACRIMSAT	December 1999	Climate variability and change
Aqua	May 2002	Atmospheric composition; carbon cycle; ecosystems; climate variability and change; water cycle
Aura	July 2004	Atmospheric composition
CALIPSO	April 2006	Atmospheric composition; water cycle
Cloudsat	April 2006	Water cycle
EO	November 2000	Carbon cycle; ecosystems
GRACE	March 2002	Water cycle; climate variability and change
ICESat	January 2003	Climate variability and change; water cycle
Jason	December 2001	Climate variability and change; water cycle
Landsat-7	April 1999	Carbon cycle; ecosystems
QuikSCAT	June 1999	Climate variability and change
SORCE	January 2003	Atmospheric composition; climate variability and change; water cycle
Terra	December 1999	Atmospheric composition; carbon cycle; ecosystems; climate variability and change; water cycle
TRMM	November 1997	Climate variability and change; water cycle

Analyses of 'historical' satellite data are combined with observations recorded from Aqua, Aura, CALIPSO, Cloudsat, and Terra, which are the primary on-orbit satellites for studies of atmospheric chemistry and composition. These missions provide data for studies of stratospheric ozone recovery, tropospheric chemistry and air quality, aerosol characterization, long-range transport of pollution, and cloud formation processes. For 2009, measurements from Aqua, Aura, CALIPSO, and Cloudsat—four of the five satellites flying in close formation in the A-Train yielding an unprecedented variety of data in a 15-minute interval on the same groundtrack—data set will be used to determine the impact of gas constituents on aerosol properties and the interactions of aerosols with clouds. Analyses of combined satellite, aircraft, and surface measurements from the Tropical Composition, Cloud, and Climate Coupling (TC4) experiment will continue to elucidate the roles of boundary layer gases and aerosols in the formation and radiative properties of cirrus clouds. The Arctic Research of the Composition of the Troposphere from Aircraft and Satellites (ARCTAS) data analyses will begin. ARCTAS will investigate effects of pollution transport, haze, halogen chemistry, and boreal forest fires on the Arctic climate. In 2009, NASA will deploy a Global Hawk unattended airborne system for calibration and validation of Aura data.

Climate Variability and Change

This research element includes ocean circulation, sea ice, ice sheets, and abrupt climate change. Feedbacks by clouds and water vapor are discussed under *Global Water Cycle*. Climate models, which have a central integrating role, are discussed under *Modeling Strategy*.

With the 2008-scheduled launch of OSTM, NASA will continue high accuracy, stable measurements of global mean sea level, which has been rising at about 3.3 mm yr^{-1} since 1993. In 2009, NASA will select a new science team to evaluate and exploit data from the Aquarius sea surface salinity mission scheduled for launch in 2010. NASA continues to lead the *U.S. Ocean Action Plan* Near-Term Opportunity to assess Atlantic meridional overturning circulation for rapid climate change.

Large changes in the Arctic and Antarctic focused scientific attention on these environments, and on the impact of these changes for the global integrated Earth system. In 2009, NASA will formulate a new satellite mission, ICESat-II, recommended in the NRC *Decadal Survey*. ICESat-II will accurately measure ice topography allowing estimation of ice sheet volume changes and sea ice thickness. An ICESat-II Science Definition Team will be established in 2009.

Global Water Cycle

The water cycle involves water in all three of its phases, exchanges large amounts of energy through phase changes, and is ubiquitous: clouds and precipitation; ocean-atmosphere, cryosphere-atmosphere, and land-atmosphere interactions; mountain snow; and groundwater.

NASA researchers continue satellite data analyses to quantify and monitor the processes, which comprise the Earth's hydrologic cycle and its variability in space and time. In 2009, investigations will focus on the components of the global water cycle (e.g., precipitation pattern) in order to examine possible changes that may occur as atmospheric water storage becomes larger with increased global warming.

In 2009, NASA will initiate a new research opportunity to better understand tropical cyclone genesis and intensification processes through the combination of satellite observations, model simulations, and airborne field campaigns. Research during the next 5 years will focus on understanding the relative roles of different rapid intensification processes ranging in scale from the synoptic environment to the convective scale.

In 2009, NASA will start the *Decadal Survey* mission SMAP to measure soil moisture and determine the freeze/thaw state of the soil when launched in 2012. SMAP data will contribute to determining water fluxes between the atmosphere, the land surface, and the subsurface of the land; extending the capabilities of medium-range weather forecast models and seasonal climate models; and understanding whether the soil is a carbon sink or source in cold regions.

Land-Use and Land-Cover Change

Land-use dynamics depends on understanding past and present land-use practices. Land-use and land-cover change drive climate change, and climate change itself drives land use and land cover.

In 2009, NASA and USGS will release a new global land survey (GLS) data set produced from Landsat-5 and Landsat-7 images recorded during 2004-2007. Land-cover changes over 30 years will be determined from comparative analyses of the new GLS data with previous GLS data sets constructed for 1975, 1990, and 2000. In 2009, NASA will support the NRC in a study of the state of knowledge and future research needs for land change modeling approaches. NASA will continue to support scientific investigations on land-use and land-cover change impacts on ecosystems in the Northern Eurasia Partnership Initiative.

Global Carbon Cycle

Research on the global carbon cycle addresses the distribution and cycling of carbon among the terrestrial, oceanic, and atmospheric reservoirs.

In 2009, NASA will continue analyses of ship and satellite measurements recorded in 2008 in the Southern Ocean to estimate air-sea exchange of carbon dioxide and other gases in high sea and high wind conditions. In 2009, NASA, in partnership with ONR and NSF, will develop advanced *in situ* technologies to calibrate satellite ocean color data. The new instrumentation will replace MOBY—a 15-year old marine optical buoy—and will support calibration of ocean color measurements from the on-orbit MODIS on Aqua and from VIIRS on NPP, which is scheduled for launch in 2010.

NASA will continue developing continental-scale data products for the North America Carbon Program (NACP) and continue modeling and synthesis activities for the NACP mid-continent intensive field campaign. In 2009, research will continue on carbon storage and emissions in northern high latitudes. In 2009, research will be initiated to evaluate atmospheric carbon dioxide measurements from the OCO satellite mission scheduled for launch in December 2008.

Ecosystems

Ecosystems affect the climate system through large exchanges of greenhouse gases. The goals of this research element are to document and understand how terrestrial and marine ecosystems are changing and to quantify carbon budgets for key ecosystems.

In 2009, NASA will combine satellite and *in situ* observations with Earth system and ecological models. New studies include those that relate land surface warming to phytoplankton productivity, characterize climate change affects on bird habitats and bird survival, and address terrestrial ecosystem response in high latitudes to climate change. In 2009, NASA will develop algorithms to analyze satellite observations of three-dimensional vegetation structure in support of *Decadal Survey* missions DESDynI and ICESat-II. Research will continue with multi-sensor observations of biological impacts of a changing climate on coral reefs.

Human Contributions and Responses

In 2009, NASA will support the evaluation and assessment of potential policy-, market-, and technology-based approaches to adapting to or mitigating the impacts of climate change. Also, NASA will address the impact of climate change to exposure and severity of infectious diseases.

Modeling Strategy

Our quantitative understanding of Earth system processes and feedbacks is codified in climate models. Climate models are rapidly improving in accuracy and sensitivity through incorporation of new understandings of the physics, chemistry, and biology and their interactions. New infrastructure and advanced computing capacity and computing capabilities are an important component of improved climate models.

The Earth System Modeling Framework (ESMF)—which was initiated in 2002 by NASA and now is an interagency activity—enables shared infrastructure and interoperability of model components and interface. In 2009, optimized data transfer, hierarchical attributes, regridding interface, and multi-tile grids will be completed. In 2009, ESMF will demonstrate interfaces for coupling ESM components on different computers.

During 2009, NASA will enhance its Earth System Model (ESM) by incorporating chemistry-physics coupling throughout the atmosphere. The ESM will include a non-hydrostatic finite volume dynamical core allowing evaluation across temporal scales from weather to climate.

Appendix A

Decision Support Resources Development

NASA will demonstrate the utility of NASA research capabilities for decision-support needs over many sectors likely to be impacted by climate change, such as agriculture, air quality, disaster management, ecosystems, public health, water resources, and weather. The focus in 2009 will be on applications to policy and decisions the Nation will consider in responding to climate change (e.g., policy frameworks, technology and market approaches, decisionmaking on national and regional climate adaptation and mitigation, and carbon and energy management).

Observing and Monitoring the Climate System

Global measurements to understand the physical, biological, chemical, and ecosystem processes responsible for changes in the Earth system on all relevant space and time scales are critical to understand past and present climate changes and predict future climate change. NASA is the world's leading organization recording sustained, high accuracy, stable, global observations with high spatial and temporal resolutions for studies of global and regional climate change.

The table on page 209 describes briefly the operating satellite missions and their relationship to CCSP research themes. In FY2009, NASA will launch two new missions. The OCO mission is scheduled for launch in December 2008, and will measure carbon dioxide concentrations in the atmosphere. These data will elucidate carbon dioxide sources and sinks on regional scales of about 1,000 km. The Glory mission, scheduled for launch in March 2009, will measure microphysical properties of aerosols to determine how aerosols contribute to net cooling or warming. Glory will join the A-Train to improve understanding of the interactions between aerosols and clouds and other phenomena. Also, Glory will extend the time series of total solar irradiance at the top of the atmosphere.

The June 2006 Nunn-McCurdy recertification of NPOESS de-manifested and de-scoped instruments that were intended to extend the climate data records of many important variables. In 2007, NASA and NOAA announced that the OMPS-Limb instrument would be placed on NPP set to launch in 2010 to mitigate the potential gap in ozone data to understand stratosphere ozone recovery in response to the Montreal Protocol. The OMPS-Limb instrument will provide the first vertically resolved ozone measurements. In 2008, NASA and NOAA agreed to incorporate a CERES instrument on NPP to mitigate the potential gap in nearly 30 years of Earth's radiant budget measurements. NASA is working to ensure continuity of the 29-year total solar irradiance time series with measurements by Glory, which is set to launch in 2009 to overlap with SORCE. In 2008, NASA and NOAA plan to recommend a satellite option to continue total solar irradiance measurements after those made by Glory.

NASA research satellites are typically designed for 3- to 5-year lifetimes. The primary mission objectives focus on demonstrating new measurement techniques and acquiring data to advance understanding of relatively short-term Earth system processes. Many NASA research satellite missions continue to operate well beyond their design lives, thus extending the data sets and enabling examination of longer period phenomena. In 2009, NASA will evaluate the extension of on-orbit missions that have completed their primary mission or will complete their primary mission in the 2 years following the evaluation.

Data Management and Information

NASA has led development of integrated data management and information systems to enhance interdisciplinary studies of the global Earth system. The goal is seamless, platform-independent, timely, and open access to integrated data, products, and information to address the science of climate change.

The first phase of a major upgrade to NASA's Earth Observing System Data and Information System (EOSDIS), which acquires nearly 4 terabytes of satellite data per day, was completed in 2008. Upgrades envisioned for 2009 and beyond include research on connectivity of NASA data with other relevant data sources and systems, and on seamless integration of multiple data and metadata. Research investigations continuing in 2009 will link together data from multiple satellites, create Earth system data records, and develop new data system tools that improve the access, use, and interoperability of satellite data.

NASA continues populating the Global Change Master Directory (GCMD)—a 20,000-metadata directory—at the rate of about 900 descriptions per month. A new version of GCMD software will be released in 2008 with updates planned in 2009.

NATIONAL SCIENCE FOUNDATION

Principal Areas of Focus

National Science Foundation (NSF) programs address global change issues through investments that advance frontiers of knowledge and provide state-of-the-art instrumentation and facilities while also cultivating a diverse highly trained workforce and developing resources for public education. In particular, NSF global change research programs support research and related activities to advance the fundamental understanding of physical, chemical, biological, and human systems and the interactions among them. The programs encourage interdisciplinary activities and focus particularly on Earth system processes and the consequences of change. NSF programs facilitate data acquisition and information management activities necessary for fundamental research on global change, and promote the enhancement of models designed to improve understanding of Earth system processes and interactions, and to develop advanced analytic methods to facilitate basic research. NSF also supports fundamental research on the processes used by organizations to identify and evaluate policies for mitigation, adaptation, and other responses to the challenge of varying environmental conditions. Through its investment, NSF contributes to the overall CCSP goals by providing a comprehensive scientific foundation for many of the synthesis and assessment products identified in the *Strategic Plan for the U.S. Climate Change Science Program*.

Program Highlights for FY 2009

Atmospheric Composition

NSF programs in tropospheric and stratospheric chemistry will continue to address the composition of the atmosphere and its relation to climate variability and change. Studies of the transformation and transport of gaseous constituents and aerosols provide insights into the radiative and cloud nucleating properties of the atmosphere. Studies of the global distributions of greenhouse gases and aerosols will provide input for future scenarios of radiative forcing.

Climate Variability and Change

NSF programs maintain a strong emphasis on climate variability and change through a combination of observational campaigns and numerous analytical and modeling activities. Ocean science efforts will focus on changes in ocean structure, circulation, and interactions with the atmosphere to improve current understanding of the processes and models that address future changes, particularly those that may happen abruptly. Studies of decadal variability and changes in the statistics of extreme weather events will be an area of emphasis in climate change research. The Community Climate System Model (CCSM) continues to be enhanced with better model physics and parameterizations as well by incorporating interactive chemistry and biology. It also is being used in combination with an embedded, state-of-the-art, mesoscale model to explore high-resolution decadal climate prediction. Studies of paleoclimatology will continue to be supported as a means to provide baseline data on natural climate variability from the past and from key climatic regions and to enable reconstructions and evaluations of past environmental change as inputs for model validations.

Global Water Cycle

NSF supports a broad-based effort to understand all aspects of the global water cycle with continued emphasis on interdisciplinary research. Relevant programs will continue to explore ways to optimally and effectively utilize the wide range of hydrological data types—continuous and discrete time and space information from a variety of platforms for research purposes. Information from process studies will be used to refine models through scaling and parameterizations of sub-grid processes, particularly the fluxes of water through the Earth system. Planning for and the initiation of several prototype hydrological observatories, both physical and virtual, are being carried out. Science and Technology Centers will continue to work with stakeholders responsible for water management and with educators to translate research advances into useful products, particularly exploring issues related to decisionmaking in the face of uncertainty as applied to the urbanizing and drought-prone Southwest. NSF's International Polar Year investments emphasize ice sheet change as a contribution to understanding the global water cycle.

Land-Use and Land-Cover Change

Several NSF programs continue to address key aspects of land-use and land-cover change through studies in ecological rates of change and related species diversity; Arctic systems; temporal variability; water and energy influences on vegetative systems; fire-land cover interactions; and diverse human influences on land utilization.

Global Carbon Cycle

NSF supports a wide variety of carbon cycle research activities. Investigations examine a range of topics in terrestrial and marine ecosystems and their relations to the carbon cycle. Research in terrestrial settings will explore, for example, carbon storage, controls on carbon exchange between ecosystems and the atmosphere, delivery of carbon by rivers, carbon fluxes from high-latitude soils, and carbon export from mountains and submarine groundwater discharge. In the oceans, air-sea gas exchange, remineralization of particles in the mesopelagic, and the upper ocean carbon budget will be addressed. Carbon cycle studies will integrate observational data into models to provide insights for understanding key aspects of the global carbon cycle.

Ecosystems

Several NSF programs support investigations of the effects of climate change on terrestrial and marine ecosystems and the consequences of those changes on climate through observational, experimental, modeling, and laboratory studies. The collection of information and knowledge of climate-ecosystem interactions in terrestrial, freshwater, and marine systems through the Long-Term Ecological Research (LTER) projects derives from the rich array of observation, monitoring, experimentation, and modeling throughout this networked research program. The Hawaii Ocean Times-Series (HOT) and Bermuda Atlantic Time Series (BATS) sites augment the LTER network in the central ocean gyre ecosystems. The Global Ocean Ecosystem Dynamics program will continue to study the impact of global ocean changes on marine ecosystems through specific synthesis activities focused on the North Atlantic, North Pacific, and Southern Ocean systems. NSF also supports studies of the production and consumption of greenhouse gases and aerosols by terrestrial and freshwater biota.

Human Contributions and Responses

NSF supports basic research on the processes through which people (individually, in groups, or through organizations) interact with natural environmental systems. Programs support projects that focus on decisionmaking under uncertainty associated with climate change. These projects are expected to produce new knowledge and tools that should facilitate improved decisionmaking by various stakeholder groups trying to deal with uncertainties associated with future climate variability and change.

International Research and Cooperation

The "International Polar Year 2007-2008" (IPY) extends from March 2007 through March 2009. The President's Office of Science and Technology Policy designated NSF the lead Federal agency in organizing U.S. IPY activities. NSF IPY activities will focus on improving understanding of climate change in both polar regions and on linkages between polar and global systems. Observational foci include stratospheric and tropospheric chemistry over Antarctica and the Southern Ocean carbon cycle. In addition, NSF, in cooperation with NASA and international partners, will focus on longer term studies that will elucidate how sea level is changing in concert with changes in the stability of the Greenland and Antarctic ice sheets.

Related Research and Education Efforts

NSF will continue to support "contributing" research on broader topics that are closely related to global and climate change. These include, *inter alia*, studies of the atmosphere, ocean, land surface, ecosystems, paleoclimatology, and human dimensions—all of which add substantively to the specific programs supporting CCSP objectives. NSF has the computing infrastructure in place and under enhancement to enable more effective utilization of the research information. In addition, NSF supports projects that integrate research with education on global and climate change to demonstrate that scientific visualization—incorporated into inquiry-based learning—can enable students to develop an understanding of complex global change phenomena. Students address these issues by evaluating multimedia data at various spatial and temporal resolutions, reviewing scientific evidence, and considering social concerns that contribute to global and climate change debates. In collaboration with other agencies, NSF is working with the geosciences education community to develop a framework for Earth System Science Literacy.

SMITHSONIAN INSTITUTION

National Air and Space Museum (NASM)
National Museum of Natural History (NMNH)
National Zoological Park (NZP)
Smithsonian Astrophysical Observatory (SAO)
Smithsonian Environmental Research Center (SERC)
Smithsonian Tropical Research Institute (STRI)

Principal Areas of Focus

Within the Smithsonian Institution, global change research is conducted at the Smithsonian Astrophysical Observatory, the National Air and Space Museum, the Smithsonian Environmental Research Center, the National Museum of Natural History, the Smithsonian Tropical Research Institute, and the National Zoological Park. Research is organized around themes of atmospheric processes, ecosystem dynamics, observing natural and anthropogenic environmental change on daily to decadal time scales, and defining longer term climate proxies present in the historical artifacts and records of the museums as well as in the geologic record at field sites. The Smithsonian Institution program strives to improve knowledge of the natural processes involved in global climate change, to provide a long-term repository of climate-relevant research materials for present and future studies, and to bring this knowledge to various audiences, ranging from scholarly to the lay public. The unique contribution of the Smithsonian Institution is a long-term perspective—for example, undertaking investigations that may require extended study before producing useful results and conducting observations on sufficiently long (e.g., decadal) time scales to resolve human-caused modification of natural variability.

Program Highlights for FY 2009

Atmospheric Composition
At SERC, measurements will be made of spectral UV-B in Maryland (>25-year record), Florida, Arizona, and other sites in the United States. These data will be electronically disseminated to meet the needs for assessing the biological and chemical impact of varying ultraviolet radiation exposures.

Climate Variability and Change
Research at NASM will emphasize the use of remote-sensing data to improve theories of drought, sand mobility, soil stability, and climate change in the Mojave Desert and Simpson Desert, Australia. Studies at NMNH and STRI will focus on the paleoecology of climate change.

Terrestrial and Marine Ecosystems
Several Smithsonian programs will examine biological responses to global change. At SERC, research will be conducted on the responses of global ecosystems to increasing carbon dioxide concentrations (also a contribution to the Global Carbon Cycle program). This SERC program will also focus on invasive species and solar UV-B. Biodiversity education and research will be performed at STRI, NMNH, and NZP. Tropical biodiversity research programs monitor global change effects through repeated sampling of flora and fauna in tropical and temperate forests, and identifying the physical and

biological processes of growth and decline of species. Other studies on ecosystem response to increasing habitat fragmentation will be conducted at NZP.

Human Dimensions of Global Change

The general public and research community will be informed of global change research conducted by Smithsonian and other CCSP agencies via exhibits. During FY 2009, an exhibition on soils developed by staff at NMNH and SERC will continue to be displayed at NMNH. Part of the "Forces of Change" series, the exhibition includes soils' role as atmospheric sources and sinks. FY 2009 will also see the debut of the new Ocean Hall (SI-NOAA joint collaboration) that deals with issues such as loss of sea ice habitat and coral reef ecosystems due to global warming and ocean acidification.

Related Research

Much of the global change research performed at the Smithsonian is not supported by direct Federal appropriation (i.e., CCSP cross-cut funding) and instead is supported by other public and private sources (including other CCSP-participating agencies). These projects are nonetheless organized around CCSP program elements, thus amplifying the scope and impact of research supported directly by CCSP. At SAO, there are extensive measurement programs for stratospheric and tropospheric composition. These include pollution measurement from space and its eventual development into continuous global monitoring. This work contributes to global climate observations, enhances climate modeling systems, quantifies greenhouse gas sources and sinks, and reduces scientific uncertainties of aerosol effects. There are continuing studies on solar activity and its relationship to climate. SERC and STRI receive agency support via competitive grants programs to perform studies of ecosystem responses to increased carbon dioxide, UV-B, and invasive species. Other contributing activities include research conducted by several units within the Smithsonian in a variety of habitats concerning natural and human-induced variations in species, populations-communities, and ecosystems. These studies help clarify the relative importance of global change effects as one of several agents of ecological change. Studies of environmental change over long time periods are aided by the Institution's collections. Used by researchers around the world, these materials provide raw data for evaluating changes in the physical and biological environment that occurred before human influences.

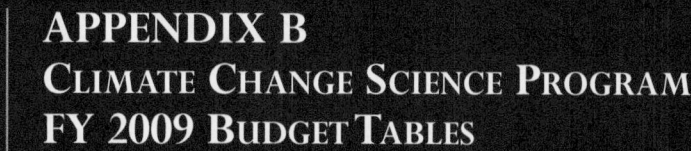

APPENDIX B
Climate Change Science Program
FY 2009 Budget Tables

APPENDIX B

CLIMATE CHANGE SCIENCE PROGRAM
FY 2009 BUDGET TABLES

CCSP integrates federally supported research on global change and climate change, as conducted by 13 U.S. Government departments and agencies:

- Department of Agriculture (USDA)
- Department of Commerce (DOC)
- Department of Defense (DOD)
- Department of Energy (DOE)
- Department of Health and Human Services (HHS)
- Department of the Interior (DOI)
- Department of State (DOS)
- Department of Transportation (DOT)
- Agency for International Development (USAID)
- Environmental Protection Agency (EPA)
- National Aeronautics and Space Administration (NASA)
- National Science Foundation (NSF)
- Smithsonian Institution (SI).

CCSP incorporates and integrates the U.S. Global Change Research Program (USGCRP) with the Administration's U.S. Climate Change Research Initiative (CCRI). CCSP budget requests are coordinated through interagency research working groups and other mechanisms, but ultimate budget accountability resides with the participating departments and agencies. As a result of its interagency composition, activities of CCSP participating agencies are funded by Congress through eight of the 12 subcommittee Appropriations bills.

The following tables summarize the CCSP budget:

- Tables 1 and 2 show the CCSP FY 2007-2009 budget aligned by CCSP goal.
- Table 3 shows the CCSP FY 2007-2009 budget by agency, with USGCRP and CCRI budgets listed separately and also combined in a single CCSP total.
- Table 4 shows the USGCRP FY 2007-2009 budget by CCSP research element.
- Table 5 shows the FY 2007-2009 CCRI budget.
- Subsequent tables show, for each CCSP participating agency, the FY 2007-2009 budget for both USGCRP and CCRI activities.

TABLE 1. FY 2007–2009 CLIMATE CHANGE SCIENCE PROGRAM
BUDGET BY GOAL AND FOCUS AREA

Focus Area	Description (from *CCSP Strategic Plan*)[1]	**Budgets ($M)[2]** FY 2007	FY 2008 Estimate	FY 2009 Request	Agencies
Goal 1	**Improve knowledge of the Earth's past and present climate and environment, including its natural variability, and improve understanding of the causes of observed variability and changes**				
Focus 1.1	Better understand natural long-term cycles in climate [e.g., Pacific Decadal Variability (PDV), North Atlantic Oscillation (NAO)]	39.6	43.6	47.3	DOC, DOE, DOI, NASA, NSF
Focus 1.2	Improve and harness the capability to forecast El Niño-La Niña and other seasonal-to-interannual cycles of variability	37.0	35.4	37.1	DOC, DOE, DOI, NASA, NSF
Focus 1.3	Sharpen understanding of climate extremes through improved observations, analysis, and modeling, and determine whether any changes in their frequency or intensity lie outside the range of natural variability	35.8	37.0	42.0	DOC, DOE, DOI, NASA, NSF
Focus 1.4	Increase confidence in the understanding of how and why climate has changed	38.4	39.2	43.8	DOE, DOI, NASA, NSF, SI
Focus 1.5	Expand observations and data/information system capabilities	173.7	191.1	240.4	DOC, DOE, DOI, NASA, NSF, SI
GOAL 1 TOTAL		**324.5**	**346.3**	**410.6**	
Goal 2	**Improve quantification of the forces bringing about changes in the Earth's climate and related systems**				
Focus 2.1	Reduce uncertainties about the sources and sinks of greenhouse gases, emissions of aerosols and their precursors, and their climate effects	94.1	96.2	103.9	DOC, DOE, DOI, DOT, NASA, NSF
Focus 2.2	Monitor the recovery of the ozone layer and improve the understanding of the interactions of climate change, ozone depletion, tropospheric pollution, and other atmospheric issues	27.3	28.1	30.8	USDA, DOE, NASA
Focus 2.3	Increase knowledge of the interactions among emissions, long-range atmospheric transport, and transformations of atmospheric pollutants, and their response to air quality management strategies	39.1	40.4	43.0	NASA, NSF

[1] See Chapter 2 of the *Strategic Plan for the U.S. Climate Change Science Program* for a detailed discussion.
[2] Any minor discrepancies within this table and between this table and others are due to rounding.

TABLE 1 (CONTINUED)

Focus Area	Description (from *CCSP Strategic Plan*)[1]	Budgets ($M)[2]			Agencies
		FY 2007	FY 2008 Estimate	FY 2009 Request	
Goal 2 (continued)					
Focus 2.4	Develop information on the carbon cycle, land cover and use, and biological/ecological processes by helping to quantify net emissions of carbon dioxide, methane, and other greenhouse gases, thereby improving the evaluation of carbon sequestration strategies and alternative response options	127.6	132.0	134.2	USDA, DOC, DOE, DOI, NASA, NSF, SI
Focus 2.5	Improve capabilities to develop and apply emissions and related scenarios for conducting "If…, then…" analyses in cooperation with CCTP	3.0	3.0	3.0	DOE
GOAL 2 TOTAL		**291.0**	**299.6**	**314.8**	
Goal 3	**Reduce uncertainty in projections of how the Earth's climate and related systems may change in the future**				
Focus 3.1	Improve characterization of the circulation of the atmosphere and oceans and their interactions through fluxes of energy and materials	37.6	38.5	44.4	DOC, DOE, DOI, NASA, NSF
Focus 3.2	Improve understanding of key "feedbacks" including changes in the amount and distribution of water vapor, extent of ice and the Earth's reflectivity, cloud properties, and biological and ecological systems	66.0	66.8	69.4	DOE, DOI, NASA, NSF
Focus 3.3	Increase understanding of the conditions that could give rise to events such as rapid changes in ocean circulation due to changes in temperature and salinity gradients	7.5	11.8	12.6	DOE, DOI, NASA, NSF
Focus 3.4	Accelerate incorporation of improved knowledge of processes and feedbacks into climate models to reduce uncertainty in projections of climate sensitivity, changes in climate, and related conditions such as sea level	84.1	89.8	103.0	DOC, DOE, NASA, NSF
Focus 3.5	Improve national capacity to develop and apply climate models	41.8	43.3	50.6	DOC, DOE, NASA, NSF
GOAL 3 TOTAL		**236.9**	**250.1**	**279.8**	

TABLE 1 (CONTINUED)

Focus Area	Description (from *CCSP Strategic Plan*)[1]	Budgets ($M)[2]			Agencies
		FY 2007	FY 2008 Estimate	FY 2009 Request	
Goal 4	**Understand the sensitivity and adaptability of different natural and managed ecosystems and human systems to climate and related global changes**				
Focus 4.1	Improve knowledge of the sensitivity of ecosystems and economic sectors to global climate variability and change	62.5	60.8	62.8	USDA, DOE, DOI, DOT, EPA, NASA, NSF, SI
Focus 4.2	Identify and provide scientific inputs for evaluating adaptation options, in cooperation with mission-oriented agencies and other resource managers	56.5	57.9	57.5	HHS, DOI, DOT, EPA, NSF
Focus 4.3	Improve understanding of how changes in ecosystems (including managed ecosystems such as croplands) and human infrastructure interact over long time periods	40.1	43.1	39.7	USDA, DOC, DOI, DOT, NASA, NSF, SI
GOAL 4 TOTAL		**159.1**	**161.8**	**160.0**	
Goal 5	**Explore the uses and identify the limits of evolving knowledge to manage risks and opportunities related to climate variability and change**				
Focus 5.1	Support informed public discussion of issues of particular importance to U.S. decisions by conducting research and providing scientific synthesis and assessment reports	57.9	52.2	52.8	USDA, DOI, EPA, NASA, NSF, SI
Focus 5.2	Support adaptive management and planning for resources and physical infrastructure sensitive to climate variability and change; build new partnerships with public and private sector entities that can benefit both research and decisionmaking	62.0	66.1	72.0	USDA, DOC, DOI, USAID, EPA, NASA, NSF
Focus 5.3	Support policymaking by conducting comparative analyses and evaluations of the socioeconomic and environmental consequences of response options	18.4	20.8	19.0	USDA, DOE, DOI, EPA, NASA, NSF, SI
GOAL 5 TOTAL		**138.3**	**139.1**	**143.8**	
CCSP PROGRAM TOTAL		**1,149.8**	**1,197.0**	**1,309.0**	

TABLE 2. FY 2007-2009 CLIMATE CHANGE SCIENCE PROGRAM
BUDGET BY GOAL AND PARTICIPATING AGENCY/DEPARTMENT

[DISCRETIONARY BUDGET AUTHORITY IN $M]

	USDA	DOC[1]	DOE	HHS	DOI	DOT	USAID	EPA	NASA	NSF	SI	Res Subtotal	Obs[2] Subtotal	TOTAL
						Research								
FY 2009 Request														
Goal 1	0.0	217.7	47.3	0.0	11.9	0.0	0.0	0.0	80.8	51.8	1.1	410.6	399.6	810.2
Goal 2	19.6	41.6	33.1	0.0	4.5	1.9	0.0	0.0	149.6	64.1	0.4	314.8	127.9	442.7
Goal 3	0.0	46.5	50.4	0.0	1.2	0.0	0.0	0.0	127.0	54.7	0.0	279.8	187.3	467.1
Goal 4	16.8	2.0	13.1	46.8	12.2	0.0	0.0	3.6	21.3	41.0	3.2	160.0	53.4	213.4
Goal 5	23.9	16.8	2.0	0.0	1.6	0.0	20.0	12.7	56.8	9.0	1.0	143.8	0.0	143.8
TOTAL	60.3	324.6	145.9	46.8	31.4	1.9	20.0	16.3	435.5	220.6	5.7	**1,309.0**	768.2	**2,077.2**
FY 2008 Estimate														
Goal 1	0.0	162.0	46.4	0.0	12.0	0.0	0.0	0.0	75.3	49.5	1.1	346.3	341.1	687.4
Goal 2	19.3	38.7	32.2	0.0	5.3	0.5	0.0	0.0	143.2	60.0	0.4	299.6	119.7	419.3
Goal 3	0.0	46.1	34.8	0.0	1.3	0.0	0.0	0.0	119.2	48.7	0.0	250.1	163.6	413.7
Goal 4	18.5	4.7	13.1	46.8	12.3	0.0	0.0	4.4	20.8	38.0	3.2	161.8	41.8	203.6
Goal 5	26.7	14.8	2.0	0.0	2.6	0.0	14.0	15.2	53.8	9.0	1.0	139.1	0.0	139.1
TOTAL	64.5	266.3	128.5	46.8	33.5	0.5	14.0	19.6	412.3	205.3	5.7	**1,197.0**	666.2	**1,863.2**
FY 2007														
Goal 1	0.0	139.7	44.6	0.0	11.0	0.0	0.0	0.0	78.6	49.5	1.1	324.5	298.9	623.4
Goal 2	16.9	37.6	31.0	0.0	3.1	0.4	0.0	0.0	140.2	61.4	0.4	291.0	173.9	464.9
Goal 3	0.0	42.1	29.8	0.0	1.4	0.0	0.0	0.0	114.9	48.7	0.0	236.9	163.9	400.8
Goal 4	17.9	3.8	17.8	46.8	9.5	0.4	0.0	2.7	19.0	38.0	3.2	159.1	36.9	196.0
Goal 5	26.2	12.5	2.0	0.0	1.5	0.0	14.0	14.3	57.8	9.0	1.0	138.3	0.0	138.3
TOTAL	61.0	235.7	125.2	46.8	26.5	0.8	14.0	17.0	410.5	206.6	5.7	**1,149.8**	673.6	**1,823.4**

[1] The FY 2007 amount of $236 million reflects an estimate of climate expenditures. The recently calculated actual expenditures in FY 2007 were $265 milion.
[2] NASA observing systems.
Note: Any minor discrepancies within this table and between this table and others are due to rounding.

TABLE 3. CLIMATE CHANGE SCIENCE PROGRAM: FY 2007–2009 BUDGET BY AGENCY

[DISCRETIONARY BUDGET AUTHORITY IN $M]

Agency	FY 2007			FY 2008 Estimate			FY 2009 Request		
	USGCRP	CCRI	CCSP	USGCRP	CCRI	CCSP	USGCRP	CCRI	CCSP
USDA	53	8	61	57	8	65	54	8	62
DOC[1,2,3,4]	190	46	236	222	44	266	281	44	325
DOE[5]	101	25	126	103	25	128	121	25	146
HHS	47	0	47	47	0	47	47	0	47
DOI	27	0	27	34	0	34	31	0	31
DOT	0	1	1	0	1	1	0	2	2
USAID	0	14	14	0	14	14	6	14	20
EPA	16	0	16	20	0	20	16	0	16
NASA[6]	376	35	411	378	34	412	400	35	436
NSF	182	25	207	180	25	205	196	25	221
SI	6	0	6	6	0	6	6	0	6
Scientific Research Total	**998**	**154**	**1,152**	**1,047**	**151**	**1,198**	**1,158**	**153**	**1,312**
NASA Space-Based Observations	580	94	674	626	40	666	734	34	769
CCSP Total[7]	1,577	248	1,825	1,673	191	1,864			
President's Request							1,892	183	2,080

Notes:

1. NOAA previously reported its climate research activities to CCSP, which were included under its Office of Oceanic and Atmospheric Research (OAR) line office and the National Marine Fisheries Service (NMFS) line office starting in FY 2006. For FY 2008, NOAA made a decision to report activities for the NOAA climate strategic goal, as defined in the NOAA strategic plan (2005), to ensure consistent reporting and to provide the most accurate picture of its climate funding to date. The climate goal includes both research and operations funding under the following offices: OAR, NMFS, the National Weather Service, and the National Environmental Satellite, Data, and Information Service.

2. Past reports have erroneously presented all of NOAA's CCSP funding in the Operations, Research, and Facilities (ORF) account. Climate-related activities have been and continue to be funded in both the ORF account and the Procurement, Acquisition, and Construction (PAC) account.

3. DOC FY 2008 and FY 2009 funding includes new measurement and standards-related activities that NIST will undertake to support CCSP.

4. The FY 2007 amount of $236 million reflects an estimate of climate expenditures. The recently calculated actual expenditures in FY 2007 were $265 million.

5. The majority of the FY 2009 DOE increase is due to increased climate modeling efforts. Examples include testing new convection and cloud parameterization schemes, research on effects of improved initialization of coupled model components on decadal predictability of climate, and understanding the role of cryospheric processes in the climate system.

6. The NASA climate change funding levels in this table are consistent with amounts reported in the President's proposed FY 2009 budget. This table does not reflect the revised accounting approach to be instituted in FY 2009 in response to FY 2008 Consolidated Appropriations Act direction.

7. Operational space-based, surface, and in situ observing systems and programs are not included in the CCSP budget cross-cut, but contribute to achieving CCSP goals. Because DOD research activities are conducted for defense-related missions, they are not included in the CCSP budget cross-cut; however, related DOD research contributes to CCSP goals.

* Any minor discrepancies within this table and between this table and others are due to rounding.

TABLE 4. CLIMATE CHANGE SCIENCE PROGRAM:
FY 2007-2009 USGCRP SCIENTIFIC RESEARCH BUDGET BY CCSP RESEARCH ELEMENT

[DISCRETIONARY BUDGET AUTHORITY IN $M]

Agency	Atmospheric Composition	Climate Variability	Carbon Cycle	Water Cycle	Ecosystems	Land Use	Human Contributions	TOTAL
FY 2009 USGCRP Research Elements								
USDA	20.9	–	12.4	4.6	14.8	0.1	–	52.8
DOC	18.2	223.3	14.1	10.0	2.1	–	12.8	280.5
DOE	13.3	82.5	11.0	–	13.5	–	2.0	122.3
HHS	–	–	–	–	–	–	46.8	46.8
DOI	–	7.0	2.5	6.7	7.5	6.7	1.0	31.4
USAID	–	–	–	–	–	–	6.0	6.0
EPA	7.4	–	–	3.7	3.7	–	1.7	16.5
NASA	83.0	83.3	49.1	96.1	40.6	19.4	28.4	399.9
NSF	22.8	88.9	32.2	16.4	21.3	2.8	11.2	195.6
SI	0.1	1.3	0.3	–	3.2	0.8	–	5.7
TOTAL	**165.7**	**486.3**	**121.6**	**137.5**	**106.7**	**29.8**	**109.9**	**1,157.5**
FY 2008 USGCRP Research Elements								
USDA	21.8	–	12.8	5.6	15.7	0.3	–	56.2
DOC	15.9	170.2	13.8	9.9	1.5	–	10.8	222.1
DOE	12.9	65.4	10.8	–	13.4	–	2.0	104.5
HHS	–	–	–	–	–	–	46.8	46.8
DOI	–	6.9	3.5	8.1	7.5	6.6	1.0	33.6
EPA	8.8	–	–	4.4	4.4	–	2.0	19.6
NASA	78.0	79.1	47.6	89.5	39.8	19.8	24.5	378.3
NSF	20.7	78.4	30.2	16.4	20.5	2.8	11.2	180.2
SI	0.1	1.3	0.3	–	3.2	0.8	–	5.7
TOTAL	**158.2**	**401.3**	**119.0**	**133.9**	**106.0**	**30.3**	**98.3**	**1,047.0**
FY 2007 USGCRP Research Elements								
USDA	19.7	–	11.9	4.9	15.8	0.3	–	52.6
DOC	15.7	146.5	10.2	9.5	1.5	–	6.6	190.0
DOE	12.5	61.8	10.2	–	18.1	–	0.5	103.1
HHS	–	–	–	–	–	–	46.8	46.8
DOI	–	6.9	2.5	3.2	6.3	6.6	1.0	26.5
EPA	6.6	–	–	0.5	3.8	–	6.1	17.0
NASA	77.9	79.9	45.2	83.8	39.9	20.4	28.4	375.5
NSF	22.1	78.4	30.2	16.4	20.5	2.8	11.2	181.6
SI	0.1	1.3	0.3	–	3.2	0.8	–	5.7
TOTAL	**154.6**	**374.8**	**110.5**	**118.3**	**109.1**	**30.9**	**100.6**	**998.8**

Note: Any minor discrepancies within this table and between this table and others are due to rounding.

TABLE 5. FY 2007–2009 BUDGET
FOR THE CLIMATE CHANGE RESEARCH INITIATIVE (CCRI)

Agency	Program Title	FY 2007	FY 2008 Estimate	FY 2009 Request
USDA	Carbon Cycle Research (ARS)	0.7	0.6	0.2
	Carbon Cycle Research (FS)	3.6	3.5	3.6
	Carbon Management Research	2.0	1.9	2.0
	Regional and Sectoral Impacts of Climate Change	1.0	1.0	0.6
	Carbon Inventory and Analysis	1.1	1.1	1.1
U.S. Department of Agriculture CCRI Total		**8.4**	**8.1**	**7.5**
DOC	Climate Variability and Change	39.2	38.4	38.4
	Carbon Cycle	6.5	5.8	5.9
Department of Commerce CCRI Total		**45.7**	**44.2**	**44.3**
DOE	Atmospheric Radiation Measurement Program	5.6	5.6	5.6
	CCRI Climate Modeling	9.7	12.2	12.2
	CCRI Carbon Cycle	2.4	2.9	2.9
	CCRI Integrated Assessment	4.6	3.0	3.0
Department of Energy CCRI Total		**22.3**	**23.7**	**23.7**
DOT	Partnership for Air Transportation Noise and Emissions Reduction	0.4	0.3	0.3
	NextGen – Aviation Climate Change Research Initiative	0.0	0.5	1.5
	Aviation Environment Design Tool (AEDT) and System for Assessing Global Emissions (SAGE)	0.0	0.3	0.2
	DOT-wide Climate Change Center[1]	0.4	0.0	0.0
Department of Transportation CCRI Total		**0.8**	**1.0**	**2.0**
USAID	Famine Early Warning System Network (FEWS NET)	14.0	14.0	14.0
U.S. Agency for International Development CCRI Total		**14.0**	**14.0**	**14.0**
NASA Science	Atmospheric Composition	2.6	2.7	2.7
	Climate Variability	4.8	4.9	4.9
	Carbon Cycle	6.5	6.6	6.6
	Water Cycle	6.4	6.4	6.4
	Terrestrial and Marine Ecosystems	2.1	2.2	2.1
	Land-Cover/Land-Use Change	1.1	1.2	1.1
	Human Contributions and Responses	11.6	9.9	11.6
NASA Space	Atmospheric Composition	46.8	20.1	17.1
	Climate Variability	46.8	20.1	17.1
National Aeronautics and Space Administration CCRI Total[2]		**128.7**	**74.1**	**69.6**
NSF	Carbon Fluxes and Cycle	10.0	10.0	10.0
	Human Dimensions of Climate Change	5.0	5.0	5.0
	Modeling Strategy	10.0	10.0	10.0
National Science Foundation CCRI Total		**25.0**	**25.0**	**25.0**
Total Climate Change Research Initiative President's Request		**244.9**	**190.1**	**186.1**

[1] No line item exists for this expenditure. Modal agencies make voluntary contributions to the Center.
[2] NASA funding decreases for CCRI from FY 2007 to FY 2009 are primarily due to the planned ramp-down of resources for the Glory mission, a major contributor to NASA CCRI funding, which is completing development in preparation for a December 2008 (FY 2009) launch.
Note: Any minor discrepancies within this table and between this table and others are due to rounding.

U.S. DEPARTMENT OF AGRICULTURE

Program Title	FY 2007	FY 2008 Estimate	FY 2009 Request
USGCRP			
Global Carbon Cycle	**11.9**	**12.8**	**12.4**
Agricultural Research Service	3.7	3.7	3.1
Cooperative State Research, Education, and Extension Service	0.9	0.9	2.0
Economic Research Service	0.1	0.1	0.1
Forest Service	7.3	8.1	7.3
Water Cycle	**4.9**	**5.6**	**4.6**
Agricultural Research Service	3.5	3.5	3.2
Cooperative State Research, Education, and Extension Service[1]	0.0	0.0	0.0
Forest Service	1.5	2.2	1.5
Land-Use / Land-Cover Change	**0.3**	**0.3**	**0.1**
Cooperative State Research, Education, and Extension Service	0.3	0.3	0.1
Understanding Atmospheric Composition and Chemistry	**19.1**	**19.6**	**17.9**
Agricultural Research Service	19.1	19.6	17.9
Understanding Ecosystems Changes	**15.8**	**15.7**	**14.8**
Agricultural Research Service	11.8	11.0	10.7
Cooperative State Research, Education, and Extension Service	0.4	0.4	0.5
Forest Service	3.6	4.2	3.6
Understanding the Human Dimensions of Climate Change	**0.0**	**0.0**	**0.0**
Cooperative State Research, Education, and Extension Service[2]	0.0	0.0	0.0
Support the UV-B Monitoring Network	**0.0**	**1.6**	**2.4**
Cooperative State Research, Education, and Extension Service	0.0	1.6	2.4
Other National Research Initiative	**0.6**	**0.6**	**0.5**
Cooperative State Research, Education, and Extension Service	0.6	0.6	0.5
USGCRP TOTAL	**52.7**	**56.3**	**52.9**
CCRI			
Carbon Cycle Research (ARS)	0.7	0.6	0.2
Carbon Cycle Research (FS)	3.6	3.5	3.6
Carbon Management Research	2.0	1.9	2.0
Regional and Sectoral Impacts of Climate Change	1.0	1.0	0.6
Carbon Inventory and Analysis	1.1	1.1	1.1
CCRI TOTAL	**8.3**	**8.1**	**7.4**
Department of Agriculture Total President's Request	**60.9**	**64.4**	**60.3**

[1] FY 2009 request for the USGCRP Water Cycle element within CSREES was $9,000.
[2] FY 2007 and FY 2008 amounts for the USGCRP Human Dimensions element within CSREES were $39,000.

Mapping of Budget Request to Appropriations Legislation. In the Appropriations Committee reports, Department of Agriculture CCSP activities are funded under Title I–Agricultural Programs, within the ARS, CSREES Research and Education Activities, and ERS accounts; and under Title II–Conservation Programs, within the NRCS Conservation Operations account. Also in Appropriations Committee reports, U.S. Department of Agriculture CCSP activities are funded in the USDA FS section under Title II–Related Agencies, within the FS Forest Research account.

DEPARTMENT OF COMMERCE

DOC	Program Title	FY 2007	FY 2008 Estimate	FY 2009 Request
USGCRP				
NOAA	Laboratories and Cooperative Institutes	42.6	46.5	44.4
NOAA	Competitive Research Program	70.6	85.8	90.5
NOAA	Climate Data and Information	4.4	7.6	8.3
NOAA	Climate Operations	0.9	0.5	0.9
NOAA	Climate Regimes and Ecosystem Productivity	1.5	1.5	2.1
NOAA	Operational Climate Programs	70.1	80.0	129.0
NIST	Measurements and Standards for the CCSP	0.0	0.2	5.2
USGCRP TOTAL		**190.1**	**222.1**	**280.4**
CCRI				
NOAA	Competitive Research Program	45.7	44.2	44.2
CCRI TOTAL		**45.7**	**44.2**	**44.2**
Department of Commerce Total President's Request		**235.8**	**266.3**	**324.6**

Notes:
1) Starting in FY 2006, funding to DOC/NOAA's Laboratories was included as part of DOC/NOAA CCSP activities. This is a result of the evolution of NOAA's role in CCSP.
2) Prior to 2008, DOC/NOAA reported its climate research activities under its Office of Oceanic and Atmospheric Research (OAR) line office and the National Marine Fisheries Service (NMFS) line office. For FY 2008 and beyond, NOAA began reporting activities for the NOAA climate strategic goal, as defined in the NOAA Strategic Plan (2005), to ensure consistent reporting and to provide the most accurate picture of its climate funding to date. The climate goal includes both research and operations funding under the following offices: OAR, NMFS, the National Weather Service, and the National Environmental Satellite, Data, and Information Service (NESDIS).
3) Past reports have erroneously presented all of DOC/NOAA's CCSP funding in the Operations, Research, and Facilities (ORF) account. Climate-related activities have been and continue to be funded in both the ORF account and the Procurement, Acquisition, and Construction (PAC) account
4) Beginning in FY 2008, DOC includes funding for new measurement and standards-related activities that DOC/NIST undertakes to support CCSP.
5) In addition to reporting funding for NOAA's climate strategic goal for FY 2009, DOC is also reporting $74M for climate satellite sensors funded under the NESDIS PAC account.
6) The FY 2007 amount of $236 million reflects an estimate of climate expenditures. The recently calculated actual expenditures in FY 2007 were $265 million.

Mapping of Budget Request to Appropriations Legislation. In Appropriations Committee reports, funding for National Oceanic and Atmospheric Administration CCSP activities is specified in the Laboratories and Cooperative Institutes, Competitive Research Programs, Climate Operations, and Climate Data and Information lines of the Oceanic and Atmospheric Research budget; in the Climate Regimes and Ecosystem Productivity line of the National Marine Fisheries Service budget; the Data Centers and Information Services line of the National Environmental Satellite, Data, and Information Service (NESDIS) budget; and the Local Warnings and Forecasts and Central Forecast Guidance lines of the National Weather Service (NWS) budget within NOAA's Operations, Research, and Facilities account. In addition, a portion of NOAA's climate funding is found within the Procurement, Acquisition, and Construction account for NESDIS and NWS. Funding for National Institute of Standards and Technology CCSP activities is specified in the Scientific and Technical Research and Services account.

DEPARTMENT OF ENERGY

Program Title	FY 2007	FY 2008 Estimate	FY 2009 Request
USGCRP			
Climate Change Modeling	15.6	18.8	33.2
Climate Forcing	68.1	69.4	72.6
Climate Change Response	19.6	16.4	16.4
USGCRP TOTAL	103.3	104.6	122.2
CCRI			
Atmospheric Radiation Measurement Program	5.6	5.6	5.6
Climate Modeling	9.7	12.2	12.2
Carbon Cycle	2.4	2.9	2.9
Integrated Assessment Research	4.6	3.0	3.0
CCRI TOTAL	22.3	23.7	23.7
Department of Energy Total President's Request	125.6	128.3	145.9

Mapping of Budget Request to Appropriations Legislation. In the Appropriations Committee reports, Department of Energy CCSP activities are funded under Title III–Department of Energy, within the Energy Supply, Research, and Development Activities account. Also in these Appropriations Committee reports, funding for Department of Energy CCSP activities is included as part of the appropriation for Biological and Environmental Research.

DEPARTMENT OF HEALTH AND HUMAN SERVICES

HHS	Program Title	FY 2007	FY 2008 Estimate	FY 2009 Request
	USGCRP			
NCI	Health Effects of UV Radiation	31.4	31.4	31.4
NIEHS	Health Effects of UV Radiation	13.7	13.7	13.7
NIAMS	Health Effects of UV Radiation	1.7	1.7	1.7
USGCRP TOTAL		46.8	46.8	46.8
Department of Health and Human Services Total President's Request		46.8	46.8	46.8

Mapping of Budget Request to Appropriations Legislation. In the Appropriations Committee reports, Department of Health and Human Services CCSP activities are funded under the National Institutes of Health section of Title II–Department of Health and Human Services.

DEPARTMENT OF THE INTERIOR

DOI	Program Title	FY 2007	FY 2008 Estimate	FY 2009 Request
	USGCRP			
USGS	Earth Surface Dynamics	10.5	10.3	10.5
USGS	Hydroclimatology and Water, Energy, and Biogeochemical Budgets	3.2	3.1	3.1
USGS	Land Characterization Research and Applications	2.9	2.9	2.9
USGS	Satellite Data Management and Dissemination	3.7	3.7	3.7
USGS	Terrestrial and Coastal Ecosystem Changes	6.2	6.1	6.2
USGS	Global Change Initiative	0.0	7.4	5.0
USGCRP TOTAL		26.5	33.5	31.4
Department of the Interior Total President's Request		26.5	33.5	31.4

Mapping of Budget Request to Appropriations Legislation. In the Appropriations Committee reports, Department of the Interior CCSP activities are funded under Title I–Department of the Interior. Funding for U.S. Geological Survey CCSP programs is included within the USGS Survey, Investigations, and Research account.

DEPARTMENT OF TRANSPORTATION

Program Title	FY 2007	FY 2008 Estimate	FY 2009 Request
CCRI			
Partnership for Air Transportation Noise and Emissions Reduction (PARTNER)	0.4	0.25	0.3
NextGen- Aviation Climate Change Research Initiative	0.0	0.5	1.5
Aviation Environmental Design Tool (AEDT) and			
System for Assessing Global Emissions (SAGE)	0.0	0.25	0.2
DOT-wide Climate Change Center	0.4	0.0	0.0
CCRI TOTAL	**0.8**	**1.0**	**1.9**
Department of Transportation Total President's Request	**0.8**	**1.0**	**1.9**

Note:
1) Modal administrations within the Department of Transportation contributed financially to the DOT Center for Climate Change and Environmental Forecasting in amounts less than $100,000.

Mapping of Budget Request to Appropriations Legislation. Since 2000, the Department's climate change research has been funded by contributions from eight of DOT's operating administrations and the Office of the Secretary.

U.S. AGENCY FOR INTERNATIONAL DEVELOPMENT

Program Title	FY 2007	FY 2008 Estimate	FY 2009 Request
USGCRP			
Climate Change and Energy Security Initiative	0.0	0.0	6.0
USGCRP TOTAL	0.0	0.0	6.0
CCRI			
Famine Early Warning System Network (FEWS NET)	14.0	14.0	14.0
CCRI TOTAL	14.0	14.0	14.0
U.S. Agency for International Development Total President's Request	14.0	14.0	20.0

Note:
1) The CCRI FEWS NET amounts for FY 2008 and FY 2009 of $14 million have been recently revised to reflect the actual appropriation of $12 million in FY 2008 and revised request of $13.4 milion in FY 2009.

 **Mapping of Budget Request to Appropriations Legislation**. In the Appropriations Committee reports, U.S. Agency for International Development CCSP activities are funded under Title II–Bilateral Economic Assistance: United States Agency for International Development.

ENVIRONMENTAL PROTECTION AGENCY

Program Title	FY 2007	FY 2008 Estimate	FY 2009 Request
USGCRP			
Air Quality Research and Assessment	6.6	8.8	7.4
Ecosystem Research and Assessment	3.8	–	–
Human Health Research and Assessment	1.7	–	–
Water Quality Research and Assessment	0.5	–	–
Research and Assessments of the Integrated Effects of Global Change	4.4	2.0	1.7
Water Quality / Aquatic Ecosystems Research and Assessment	–	8.8	7.3
USGCRP TOTAL	17.0	19.6	16.4
Environmental Protection Agency Total President's Request	17.0	19.6	16.4

 **Mapping of Budget Request to Appropriations Legislation**. In the Appropriations Committee reports, Environmental Protection Agency CCSP activities are funded under the EPA section of Title III–Independent Agencies, within the Science and Technology account. Appropriations Committee report language may specify more directly the funding for global change research.

NATIONAL AERONAUTICS AND SPACE ADMINISTRATION

Program Title	FY 2007	FY 2008 Estimate	FY 2009 Request
USGCRP			
Atmospheric Composition	77.9	78.0	83.0
Climate Variability	79.9	79.1	83.3
Carbon Cycle	45.2	47.6	49.1
Water Cycle	83.8	89.5	96.1
Ecosystems	39.9	39.8	40.6
Land-Cover / Land-Use Change	20.4	19.8	19.4
Human Contributions and Responses	28.4	24.5	28.4
USGCRP Scientific Research Sub-Total	375.5	378.3	400.1
Atmospheric Composition	52.7	54.6	64.2
Climate Variability	185.8	166.0	148.2
Carbon Cycle	142.1	113.5	130.4
Water Cycle	113.7	177.4	256.4
Ecosystems	42.2	52.6	62.0
Land-Cover / Land-Use Change	43.5	61.8	73.2
USGCRP Space-Based Observations Sub-Total	580.0	625.9	734.4
USGCRP TOTAL	**955.5**	**1,004.2**	**1,134.5**
CCRI			
Atmospheric Composition	2.6	2.7	2.7
Climate Variability	4.8	4.9	4.9
Carbon Cycle	6.5	6.6	6.6
Water Cycle	6.4	6.4	6.4
Ecosystems	2.1	2.2	2.1
Land-Cover / Land-Use Change	1.1	1.2	1.1
Human Contributions and Responses	11.6	9.9	11.6
CCRI Scientific Research Sub-Total	35.2	33.9	35.4
Atmospheric Composition	46.8	20.1	17.1
Climate Variability	46.8	20.1	17.1
CCRI Space-Based Observations Sub-Total	93.6	40.3	34.1
CCRI TOTAL[1]	**128.7**	**74.2**	**69.5**
National Aeronautics and Space Administration Total President's Request	**1,084.2**	**1,078.4**	**1,204.0**

[1] NASA funding decreases for CCRI from FY 2007 to FY 2009 are primarily due to the planned ramp-down of resources for the Glory mission, a major contributor to NASA CCRI funding, which is completing development in preparation for a December 2008 (FY 2009) launch.

Mapping of Budget Request to Appropriations Legislation. In the Appropriations Committee reports, National Aeronautics and Space Administration CCSP activities are funded under NASA Earth science and technology programs within Title III–Independent Agencies, as part of the Science, Aeronautics, and Technology account.

NATIONAL SCIENCE FOUNDATION

Program Title	FY 2007	FY 2008 Estimate	FY 2009 Request
USGCRP			
Atmospheric Composition	22.1	20.7	22.8
Climate Variability and Change	78.4	78.4	88.9
Carbon Cycle	30.2	30.2	32.2
Water Cycle	16.4	16.4	16.4
Terrestrial and Marine Ecosystems	20.5	20.5	21.3
Land Use / Land Cover	2.8	2.8	2.8
Human Dimensions of Climate Change[1]	11.2	11.2	11.2
USGCRP TOTAL	**181.6**	**180.3**	**195.6**
CCRI			
Carbon Fluxes and Cycle	10.0	10.0	10.0
Human Dimensions of Climate Change	5.0	5.0	5.0
Modeling Strategy	10.0	10.0	10.0
CCRI TOTAL	**25.0**	**25.0**	**25.0**
National Science Foundation Total President's Request	**206.6**	**205.3**	**220.6**

[1] NSF characterizes funding for the CCSP program "Human Dimensions of Climate Change" as "Human Contributions and Responses to Climate Change."

Mapping of Budget Request to Appropriations Legislation. In the Appropriations Committee reports, National Science Foundation CCSP activities are supported under the NSF section of Title III–Independent Agencies within the NSF Research and Related Expenses account.

SMITHSONIAN INSTITUTION

SI	Program Title	FY 2007	FY 2008 Estimate	FY 2009 Request
USGCRP				
NMNH	Archaebiology Program (Human Ecology History)	0.3	0.3	0.3
NMNH	Paleoecological Effects of Climate Change, including Evolution of Terrestrial Ecosystems	0.9	0.9	0.9
NMNH	Global Volcanism Program	0.2	0.2	0.2
NMNH	Human Origins Program (Human Ecological History)	0.3	0.3	0.3
NMNH	Nile Delta Subsidences / Sea-Level Rise	0.2	0.2	0.2
NMNH	Tropical Biodiversity Program	0.6	0.6	0.6
SERC	Ecological Effects of Ultraviolet Radiation	0.2	0.2	0.2
SERC	Effects of Increasing Atmospheric CO_2 on Ecosystems	0.3	0.3	0.3
STRI	Temperate and Tropical Forest Canopy Biology	0.4	0.4	0.4
STRI	Tropical Forest Science	0.9	0.9	0.9
STRI	Biodynamics of Forest Fragments	0.1	0.1	0.1
STRI	Tropical Agroforestry	0.2	0.2	0.2
NZP	Migratory Birds	0.3	0.3	0.3
NZP	Predicting Species Responses	0.7	0.7	0.7
NZP	Monitoring and Assessment of Biodiversity (MAB) Program	0.1	0.1	0.1
USGCRP TOTAL		**5.7**	**5.7**	**5.7**
Smithsonian Institution Total President's Request		**5.7**	**5.7**	**5.7**

Mapping of Budget Request to Appropriations Legislation. In the Appropriations Committee reports, Smithsonian Institution CCSP activities are funded in the Smithsonian section of Title II–Related Agencies, within the Salaries and Expenses account. Appropriations Committee reports specify funding for a Sciences line item component of this account, which includes CCSP programs.

CONTACT INFORMATION

Climate Change Science Program Office
1717 Pennsylvania Avenue, NW
Suite 250
Washington, DC 20006
202-223-6262 (voice)
202-223-3065 (fax)
information@climatescience.gov
information@usgcrp.gov
http://www.climatescience.gov/
http://www.usgcrp.gov/

The Climate Change Science Program incorporates the
U.S. Global Change Research Program and the Climate Change Research Initiative.

Special thanks to NASA Earth Observatory for many of the non-captioned data products used
throughout this edition of Our Changing Planet [<earthobservatory.nasa.gov>].

To obtain a copy of this document, place an order at the Global Change Research
Information Office (GCRIO) web site: http://www.gcrio.org/orders.

www.ingramcontent.com/pod-product-compliance
Lightning Source LLC
Chambersburg PA
CBHW080636180526
45168CB00008B/3189